Getting Science Wrong

ALSO AVAILABLE FROM BLOOMSBURY

The Bloomsbury Companion to the Philosophy of Science, edited by Steven French and Juha Saatsi

The History and Philosophy of Science: A Reader, edited by Daniel J. McKaughan and Holly Vande Wall

Philosophy of Science: Key Concepts, Steven French

Thomas Kuhn's Revolutions, James A. Marcum

Getting Science Wrong

Why the philosophy of science matters

PAUL DICKEN

Bloomsbury Academic
An imprint of Bloomsbury Publishing Plc

B L O O M S B U R Y
LONDON · OXFORD · NEW YORK · NEW DELHI · SYDNEY

Bloomsbury Academic

An imprint of Bloomsbury Publishing Plc

50 Bedford Square   1385 Broadway
London   New York
WC1B 3DP   NY 10018
UK   USA

www.bloomsbury.com

BLOOMSBURY and the Diana logo are trademarks of Bloomsbury Publishing Plc

First published 2018

© Paul Dicken, 2018

Paul Dicken has asserted his right under the Copyright, Designs and Patents Act, 1988, to be identified as the Author of this work.

All rights reserved. No part of this publication may be reproduced or transmitted in any form or by any means, electronic or mechanical, including photocopying, recording, or any information storage or retrieval system, without prior permission in writing from the publishers.

No responsibility for loss caused to any individual or organization acting on or refraining from action as a result of the material in this publication can be accepted by Bloomsbury or the author.

British Library Cataloguing-in-Publication Data

A catalogue record for this book is available from the British Library.

ISBN: HB: 978-1-3500-0727-7
       PB: 978-1-3500-0728-4
       ePDF: 978-1-3500-0730-7
       ePub: 978-1-3500-0729-1

Library of Congress Cataloging-in-Publication Data

A catalog record for this book is available from the Library of Congress.

Cover design: Irene Martinez Costa
Cover image © PRISMA ARCHIVO/Alamy Stock Photo

Typeset by Fakenham Prepress Solutions, Fakenham, Norfolk NR21 8NN
Printed and bound in Great Britain

To find out more about our authors and books visit www.bloomsbury.com. Here you will find extracts, author interviews, details of forthcoming events and the option to sign up for our newsletters.

Contents

List of figures vi
Introduction vii

1 Learning from our mistakes 1
2 A matter of trial and error 25
3 Images of science 47
4 88.6 percent of all statistics are made up 69
5 Living in different worlds 93
6 The bankruptcy of science 119
7 Deus ex machina 141

Epilogue 165
Dramatis personae 173
Notes 182
Bibliography 197
Index 200

List of figures

Figure 1.1 "M. Adams découvrant la nouvelle planète dans le rapport de M. Leverrier" (Mr Adams discovers the new planet in the work of Mr Leverrier) by Cham (Amedee Charles Henri de Noe).
From *L'Illustration* (November 1846)   16

Figure 2.1 "A medieval missionary tells that he has found the point where Heaven and Earth meet" (anonymous). From Camille Flammarion, *L'atmosphère: météorologie populaire* (Paris, 1888)   36

Figure 4.1 "Holmes pulled out his watch" by Sidney Paget. From *The Greek Interpreter* by Arthur Conan Doyle, published in *The Strand Magazine* (September 1893)   73

Figure 5.1 "They spoke by the clicking or scraping of huge paws!" by Howard V. Brown. From *The Shadow Out of Time* by H. P. Lovecraft, published in *Astounding Stories* (June 1936)   113

Figure 6.1 "The Temptation of St Anthony in the Desert" by Lucas Cranach. Woodcutting 1506 (Open Access Image from the Davison Art Center, Wesleyan University)   128

Introduction

This is a short book about misunderstanding science. On the face of it, this might seem like a somewhat perverse topic. Of course we understand science! After all, we rely upon the results of countless scientific theories in almost every aspect of our lives. Every time we step onto an aeroplane, visit the doctor, or eagerly shove our favorite Jean-Claude Van Damme movie into the DVD player, we are implicitly endorsing the truth of a whole range of theoretical assumptions about the world around us. Even during the course of the relatively low-tech and haphazard production of this book, I relied continually upon the outcome of countless scientific advances that would have been unimaginable only a few decades before—everything from ignoring the increasingly agitated emails from my editor and googling references online, to arguing with the word processor which checks my spelling, corrects my grammar, and otherwise brutalizes my carefully crafted prose. If we really misunderstood science and its laws, none of this would be possible. Our aeroplanes would crash, our medicines would poison us, and we would never get to see that peculiar brand of tight-trousered spin-kicking martial arts justice that we all know and love. Perhaps more pointedly, if electrons did not really exist and behave in more-or-less the way that our scientific theories say they do, then my laptop would be nothing more than an expensive paperweight, this book would never have been written, and we would not even be having this conversation in the first place.

On the other hand though, there may also be grounds for a considerably more cautious attitude. Our contemporary scientific theories are undoubtedly highly successful in their particular realm of application—but as they say, it is a big old world out there, and we have only really just begun to scratch the surface. From a cosmic perspective, the number and the range of incidents in which we have tested even our most successful scientific theories is very, very small indeed. To make matters even worse, one of the most important contributions made by modern science has been to explore just how intellectually uninspiring we are as a species, and how astronomically parochial our place in the universe. Successive advances in physics have gradually displaced the Earth from the unmoving center of a divinely ordered cosmos, to just one of many rapidly orbiting planets in a distant corner of an unremarkable galaxy. Similarly, progress in biology has only served to underline the severe cognitive limitations of a species originally evolved to hunt large mammals across a well-lit and relatively featureless plain, now struggling to comprehend such mind-boggling phenomena as the relativistic curvature of spacetime and the indeterminacy of a quantum superposition. Paradoxically enough, it would seem then that the more science tells us about the world around us, the more skeptical we should be of our ability to fully grasp it! And indeed, the history of science certainly attests to the myriad false starts and dead ends that have plagued our intellectual development. We may well feel that the indisputable success of our contemporary scientific theories gives us compelling reasons to believe them to be true—but then so too did countless scientists before us, right up to the point when the next big breakthrough sent everyone scurrying back to the chalkboard. We have been wrong before, and we will surely be wrong again.

These two competing intuitions actually provide the starting point for an entire academic subdiscipline known as the philosophy of science, a subject that I—in retrospect, perhaps unwisely—spent years teaching throughout the scattered corners of the globe. It is certainly not the most fashionable of research fields, even among other philosophers, and I can assure you that there definitely isn't very much money in it. The central questions of the philosophy of science are not ones that much concern the practicing scientist, who quite rightly has far more important things to worry about than these abstract speculations, and the most celebrated figures in the discipline are not exactly household names. In fact, now that I think about it, I probably only really got into the subject in the first place because it meant that I didn't have to take any more tedious classes in ethics. But I soon became gripped. For whatever else one may think about it, the natural sciences remain our most important method for finding out about the world around us, and if we cannot resolve this underlying tension—between the seemingly undeniable success of our contemporary scientific theories, and the equally undeniable track-record of scientific failure—then this raises some serious questions about the rest of our cognitive endeavors, and how we should think about our place in the world more generally.

All of this fascinating philosophical speculation however presupposes another, more fundamental question. In the space of the previous three paragraphs I have spoken very casually about the success of our scientific theories and the historical track-record of our scientific investigations, and have hinted at the role played by the natural sciences in shaping our understanding of ourselves and the world around us. When we argue at this level of abstraction, it is easy enough to assume that there is some well-defined notion of science to which we can uncontroversially appeal. But what exactly do we mean by science? Is there some clear set of criteria by which we can distinguish scientific practice from all of the other forms of human activity with which we might also be concerned? Or is science just bigger and better funded? How do we differentiate between good science and bad science—and those things that are only pretending to be science? Is there something about the way in which we pursue a line of enquiry that makes it scientific, or does it just depend upon which questions are currently considered to be more important than others? If science really is our most important way of finding out about the world, then it would be nice to know if there is in fact some particular way of conducting an investigation and assessing its results that is both peculiar to the natural sciences, and which ultimately contributes to their success.

This is the idea of the *scientific method*, and it is one that has obsessed many philosophers and scientists, and produced a great deal of argument and debate. But it is also an idea that has important consequences that reach beyond the rather narrow issues in the philosophy of science discussed above. For if there really is some specific way of proceeding shared by all of our most successful scientific activities, then it stands to reason that this is something that we need to examine, understand, and ultimately try to export to as many other aspects of our intellectual lives as possible. On this view, the natural sciences do not merely give us an accurate theoretical description of the world around us—they also provide us with the gold standard against which we can measure all other types of thinking. And indeed, this is something that we see creeping into most aspects of our day-to-day lives. We see the virtues of scientific thinking claimed in everything from high-level politics and policymaking, to daily squabbles on the street and the unrestrained vitriol of social media. My shampoo has apparently been "scientifically formulated" to retain the color and balance of my hair, and my breakfast yogurt is so clinically exemplary that all the different microbes and ingredients have been given special, made-up names to make them sound more important. Enrol in a modern university, and you will no longer study politics or sociology as in the bad old days, but various flavors of "political science"—along with "library science," "mortuary science," or "dairy science" depending upon the institution in question—which lets you know that what you are doing *must* be considerably more rigorous than it was before. And there is no better way of shutting up an opponent or closing down a dissenting point of view than to dismiss it as "unscientific." It is the ultimate trump card against which there can be no come back.

The problem however is that there does not appear to be any readily agreed upon account regarding the details of this supposed scientific method, in the sense of a definitive set of rules and procedures common to our most successful scientific activities and sufficient to differentiate them from all other intellectual enterprises. The suspicion is that the natural sciences enjoy such a privileged status, not because they provide an alternative to our less glamorous day-to-day methods of reasoning, but merely because they exemplify the most rigorous and precise application of those methods. It is a difference of degree, not of kind. It has taken philosophers many years to come to this realization—the wheels on this particular wagon fell off sometime in the 1970s—but it has done nothing to undermine the still popular idea that there are some objective principles of reasoning against which we can judge and subsequently use to silence those whom we do not consider to have made the grade. That is a serious idea with significant consequences; it is also one that is almost certainly wrong. That is the sort of misunderstanding of science that this book is about, and in the following chapters I will do my best to outline, criticize, and, where necessary, ruthlessly mock, some of the more widespread misconceptions regarding the scientific method that still seem to enjoy common currency today.

INTRODUCTION

*

On first inspection, it might seem fairly straightforward to specify what exactly we mean by the scientific method. We begin by making simple and unprejudiced observations of the world around us, form a tentative conjecture on the basis of those observations, and then critically test our nascent scientific theory through rigorous experimentation. If the theory fails our tests, it is abandoned; if it survives, we use it to explain the phenomena we initially observed, and to help us construct more elaborate conjectures which are then tested in turn. And at one level this is of course correct—scientists do make observations, form conjectures, and test them. But this is hardly sufficient to differentiate scientific practice from pretty much every other form of human endeavor, let alone to explain its unparalleled success and prestige.

Moreover, once we descend to a more detailed level of analysis, there emerge significant difficulties with each of these individual stages of investigation. This book begins with the notion of critical testing, and the popular idea that what distinguishes genuine scientific practice from other human activities is its willingness to put its most cherished scientific theories to the most strenuous of tests, and to readily abandon them in the light of any evidence against them. It is often noted for instance that there are any number of less intellectually respectable enterprises—astrology, homeopathy, psychoanalysis—that tend by contrast to constantly find ways to twist and amend their positions in order to accommodate any apparent anomalies. But this is only one aspect of our scientific methodology, and one that fails to capture many important instances of genuine scientific progress. More importantly, it also highlights the dangers of attempting to reduce the scientific method to a single simple set of rules or principles. This is particularly well illustrated by recent court decisions in the U.S. which have attempted to legislate against religiously inspired alternatives to the theory of evolution on the basis of such a monolithic understanding of scientific practice. The result has been not only an unqualified failure, but has also moved very close to precisely the kind of draconian narrow-mindedness that such appeals to the critical attitude of science was supposed to oppose.

This particular incident sets up several of the themes that will be explored in the rest of the book. The intention is not only to discuss some of the more common misconceptions concerning the scientific method, but also how these misconceptions have in turn influenced other aspects of our lives. The attempt to use a particular definition of the scientific method to legislate against various forms of creationism and intelligent design also indicates the way in which discussions of science are frequently colored by political motivations. The second chapter therefore examines one of the most infamous incidents in the history of science, the prosecution of Galileo by the Roman Catholic Church. This concerns another way in which we might try to individuate the scientific method, this time in terms of its reliance upon unprejudiced observation, in contrast to the sort of dogmatic appeal to authority that supposedly motivated Galileo's opponents. Unfortunately, the facts of the case turn out to be very different from the way in which they are usually perceived—unprejudiced observation turns out to be much more difficult than it seems, and political pressure upon a radical new scientific proposal can often come from *inside* the scientific community rather than from the outside world.

The third and fourth chapters concern the remaining notions of conjecture and explanation respectively. Generally speaking, our scientific theories attempt to achieve two different goals. On the one hand, they seek to provide detailed mathematical tools for predicting the behavior of different phenomena; while on the other hand, they also attempt to furnish us with satisfying explanations for why the phenomena in question behave in the way that they do. Sometimes these two different goals can pull against one another. In the case of quantum mechanics, for instance, we have a predictive tool of unprecedented precision and accuracy—but one which presents us with a view of the subatomic world that no one really understands. By contrast, the principle of natural selection in evolutionary biology gives us a powerful framework for understanding a great range of different occurrences—but does not in itself allow us to predict which mutations and adaptations are likely to survive future generations. Overemphasis on either of these two aspects leads to similar problems in understanding the scientific method. On the one hand, the ongoing revolution in computing power has raised the prospect that we can dispense with such vague and subjective notions as explanation altogether by simply reading off the relevant correlations from ever greater stores of statistical data. The problem however is that science does not just require ever more data, it also requires the *right sort* of data, and this is not something that any amount of computational number-crunching will ever be able to determine. By contrast, a closer focus on the way in which one explanation can be better than another provides us with a powerful heuristic for constructing new scientific conjectures. But there is no algorithm for determining the quality of an explanation, and no set of methodological principles that can be distilled from the process.

If the critical testing of our scientific theories fails to capture the full range of our scientific practice, if the disinterested observation of phenomena and the unprejudiced accumulation of evidence turns out to be impossible, and if our best scientific conjectures reveal more about our subjective motivations than it does the logical structure of the world, then the possibility arises that not only is there no such thing as the scientific method, but that science itself is in fact an inherently irrational activity. On this view, the central claims of our scientific theories merely reflect the larger social forces that shape the rest of our lives—better funded perhaps, and clothed with an impenetrable authority, yet still nothing more than the continuation of politics by other means. Such a view has certainly been taken up with enthusiasm by second-rate academics still yearning for their glory days of radical activism in the 1960s, but it is also a view that has dogged scientific practice since its inception. This is discussed in the fifth chapter with reference to the public reception of Einstein's theory of relativity during the post-war period—a heady mixture of faux radicalism and straightforward anti-Semitism—and how an overemphasis upon sociopolitical factors not only fails to adequately characterize scientific practice, but also leads to incoherence.

The interaction between science and society is taken up again in the sixth chapter, which considers some of the ways in which skepticism about the scientific method must similarly be handled with caution. In order to make the discussion as controversial as possible, this chapter examines some of the claims advanced in the fields of environmentalism and climate change, and in particular, some of the more apocalyptic predictions that they have produced.

It should perhaps be noted immediately that it forms no part of this book to take a stance either way on the *content* of these scientific theories—there are plenty of other books to consult, recommend, and denounce, according to one's temperament—but rather to examine the sorts of arguments that have been advanced about how we should assess these scientific theories. In particular, this chapter is concerned with how some of the skeptical arguments against e.g., man-made climate change purport somewhat incoherently to be offering broadly scientific considerations in support of their skepticism; and the way in which both sides of the debate can often conflate scientific claims about the behavior of various physical systems with sociological claims about the behavior of the individuals within those systems. Both fallacies arise, not from a misunderstanding of the scientific facts, but from an impoverished grasp of the scientific method of precisely the kind outlined in the preceding chapters.

The seventh and final chapter returns to the question of how we might differentiate science from the myriad other intellectual activities that we undertake from a broader, historical perspective. This chapter looks at some of the ways in which historians and philosophers have attempted to trace the *origins* of scientific enquiry, and to locate it in the context of the different types of narratives mankind has constructed in order to make sense of the world around it. The contrast here is usually with religion—that man becomes scientific once he stops appealing to the gods in order to explain why things occur—which is why the Darwinian revolution is so often held up as the apogee of a truly scientific world view. Yet while the idea that biological complexity can arise from simpler origins certainly provides a powerful framework for understanding the world around us, its philosophical significance is again often misunderstood and exaggerated. The realization that random variation in the face of fierce competition can lead to well-adapted outcomes is one that pre-dates Darwin by hundreds of years, which raises serious questions about the supposedly profound theological implications of contemporary biology that has fueled an entire cottage industry of popular science publications. It also threatens to prove too much. A world stripped of evidence of design may well be one liberated from all unscientific modes of thought — but it would also be one that we would not be able to comprehend, and in which our scientific methods would have no grip or purchase. Just because there is no God in the machine, it does not mean that there is no Devil in the details.

*

This book was written intermittently over the past couple of years across a number of different continents and in a variety of different circumstances. I remember that one chapter was more-or-less completed during the course of an extended train ride along the west coast of the United States, although I can't remember now if I was heading north to Seattle or south to Los Angeles, or indeed why I was taking the trip in the first place. The bar service, however, was excellent. Another chapter was written during a ferociously hot summer in Philadelphia in a combination of hipster coffee shops and spit-and-sawdust sports bars. I spent several weeks hunting down books and references through the City of Philadelphia library system for a third chapter that was eventually cut from the project altogether; nevertheless, I heartily recommend this as a strategy for getting to know your way around a new location. Most of my research on Galileo was conducted at the University of New South Wales in Sydney, and large parts of that chapter sketched out in the RSL overlooking Bondi Beach (again, great bar service). I re-read David Hume while visiting my parents back in Britain, but I only got around to putting the material together several months later during a blizzard in upstate New York, in between finishing off an undergraduate textbook that probably included far too many logical symbols. A shorter trip to Singapore was spent outlining material on medieval necromancy, which somehow managed to survive several significant redrafts and still maintains a sinister half-existence somewhere near the middle of the book. The final editing and polishing took place on the Somerset levels, for which anecdote sadly eludes me, although I was just down the road from the Burrow Hill Cider Farm, which helped a lot.

I would like to thank my agent, Andy Ross, for his enthusiasm for the project, and my editors, Colleen Coalter and Andrew Wardell for their help in seeing it to completion. This is the third book that I have written, and the first that my father has declared to be almost readable. I would like to thank him for his helpful comments; any further impugning of the reputation of Richard Nixon remains my own. Finally, my love and thanks to Katrina Gulliver for pretty much everything else.

1

Learning from our mistakes

When they successfully faked the first moon landing in July 1969, the U.S. Government managed to solve two of its most pressing political problems at a single stroke. First, of course, they managed to distract the voting public from the increasingly unpopular war in Vietnam, and thereby helped to secure the reelection of Richard Nixon, the shadowy mastermind behind the whole affair and—let's be honest about it—exactly the sort of untrustworthy character you would expect to find involved in such a cynical exploitation of the general population. Second, and perhaps more importantly, they also inflicted a humiliating defeat upon the U.S.S.R., who up until that point had clearly been winning the space race. It says a great deal about the paranoia present in the Kremlin that despite beating the Americans to every single important extra-terrestrial milestone, including the first satellite in orbit, the first animal in orbit, the first man in orbit, and even the first ever space-walk, they were nevertheless gullible enough to believe that the Apollo 11 mission could quite literally come from nowhere and trump all of their previous accomplishments.

Needless to say, it was a staggering operation and a truly spectacular feat of deception that involved the complicity of hundreds of thousands of people, from the handful of astronauts who were supposed to have walked on the moon, to the countless scientists, engineers, and ground support crew who had worked for NASA over the years and who were supposed to have made it all possible—and that's not forgetting the film crew, set designers, sound technicians, and lighting operators necessary for constructing the elaborate hoax. It involved recruiting Arthur C. Clarke and Stanley Kubrick to write and direct their iconic space-opera, *2001: A Space Odyssey*, just in order to provide the U.S. Government with a plausible cover story in case any outside agency or investigative journalist stumbled upon any of the preparations for their elaborate scam. It even involved building an enormous rocket just for the sake of appearances and blasting it into outer space, where it orbited completely undetected and quite pointedly *not* going to the moon for several days, before finally returning to Earth and splash landing in the Pacific Ocean. All in all, it would probably have been easier and certainly much cheaper to have actually just landed on the damn moon, which only goes to show just how important the whole operation must have been for the beleaguered Nixon administration, which … erm … hang on, wait a minute …

While any good conspiracy theory will of course eventually stretch its credibility well beyond breaking point, there is nevertheless something to admire amongst all the wide-eyed mutterings and tinfoil-helmet-wearing lunacy. The beauty of a good conspiracy theory is that *you can never prove it wrong*. It can accommodate absolutely any evidence you throw at it, no matter how damaging it might first appear. For example, it might occur to someone to ask, if the moon landings were indeed faked, where exactly was all that moon rock supposed to come from? But our friendly conspiracy theorist will barely even pause to glance furtively over his shoulder before explaining how all of this supposed "moon rock" was really meteor fragments dug up in the Antarctic and smuggled back to "mission control" in Houston in time for the "astronauts" to "unload" them from their "landing module" (conspiracy theories also tend to involve a lot of inverted commas). Similarly, we might point to the recent photographs taken by the Hubble telescope of the lunar landing sites as evidence of Neil Armstrong's extraterrestrial expedition. But our friendly conspiracy theorist will just shake his head sadly. Oh please. The resolution of those photos shows nothing but a few dark smudges, and anyway, everybody knows that they're all in it together. If the U.S. Government really did fake the moon landings in 1969, it stands to reason that it would also arrange for all sorts of supposedly independent evidence to support its story. In fact, it follows by a certain ruthless, internal logic that anything that you *think* proves the reality of the moon landings *must therefore* actually be part of the larger conspiracy, deliberately designed to lead you astray. In technical terms, we say that such a theory is *unfalsifiable* in the sense that it cannot ever be shown to be false—there is nothing whatsoever that anyone could ever say or do that will finally convince the conspiracy theorist that they were wrong.

All of this however stands in stark contrast to the sort of theories we normally consider within the natural sciences. Take for example something very simple, like the claim that all ravens are black. It seems easy enough to imagine the sort of evidence necessary to show that this theory is false. All it would take would be for someone to show up one morning in the laboratory with a white raven. In a situation like this, we would say that the theory had been *falsified*—again, using the word here in the purely technical sense of having been shown to be false. Or to take something a little bit more realistic, consider the basic principles of Newtonian Mechanics. This tells us that the force required to move an object is proportional to how heavy the object might be, and how fast you want it to move—or more specifically, that force equals mass times acceleration. And again, it is easy enough to imagine the sort of evidence necessary to conclusively refute such a conjecture. All it would require is an object that moved much faster than the force we put into it. There would be no comebacks, no clandestine conspiracies, no government cover-ups. In contrast to a good old fashioned conspiracy, it seems that one good counterexample is all that it takes to falsify a scientific theory once and forever.

And all of this suggests a very natural way to think about the scientific method. While any scientific theory worthy of the name will of course be based upon a careful consideration of the available evidence, the *definitive* feature of a good scientific theory is that we can easily imagine the sort of evidence necessary for showing that it was false, such as a white raven or an inexplicably fast-moving rock. In other words, the essence of the scientific method is not simply about coming up with ever more elaborate theories about the way the world works, but in subjecting those theories to the most rigorous examination possible, and then discarding those that do not make the grade. As the Austrian philosopher Karl Popper put it:

> We make progress not by amassing ever more positive evidence in favor of our scientific theories, but through the continuous elimination of one erroneous theory after another. We should think of science therefore as a constant cycle of conjecture and refutation—not of speculating about the future, but of systematically learning from our mistakes.[1]

Maybe we can never be sure that we have hit upon the correct scientific description of the world, but at least we know that as soon as that theory starts making mistakes, the scientific community will shrug their shoulders, make a fresh pot of coffee, and happily make their way back to the collective drawing board. We might not always get it right the first time around, but we make progress precisely because we have the honesty to admit when we're wrong.

This particular way of thinking about the scientific method is certainly a natural one, and has become extremely popular—not least because it dovetails neatly with some of our most cherished political convictions. Many people have argued that the development of modern science, and our intellectual emancipation from superstitious and dogmatic modes of thought more generally, owes itself to a progressively liberal political order that actively encouraged abstract speculation and the disinterested investigation of nature. But there may also be a closer connection between our scientific investigations and our political institutions. Some people have argued that they actually exemplify the *same* methodology, and can be thought of as two sides of the same coin. Popper, for example, explicitly compared the scientific method with the functioning of a democratic government. He argued that it is often difficult to know in advance which politician or political party will do the best job of running the country, and putting the matter to a vote is certainly no guarantee of selecting the best candidates. The real value of a democratic system, however, is that when we do make the wrong decision and elect the wrong people, it is easy enough to get rid of them and elect somebody else, and without having to go through the whole tiresome process of an ideological purge, raising a private army, or storming the barricades. Just as with the scientific method then, democracy comes with no guarantee that it will deliver the best result first time around—but it does come with a built-in damage control mechanism that allows us to adjust, adapt, and hopefully improve with the minimum of fuss.

For Popper then, thinking about how exactly the scientific method works is therefore about more than just uncovering a set of tools for investigating the world around us. It can also tell us something about how we see ourselves as individuals, and about how we organize ourselves as a society. That's what makes the history and philosophy of science—I think—such an exciting topic. But it also makes it a dangerous one, for while a good understanding of scientific practice can illuminate other areas of our lives in surprising and unexpected ways, a mistaken grasp of the scientific method can have disastrous implications across many different aspects of our lives. It can misinform legal decisions, misdirect social policy, justify pernicious political ideologies, and generally undermine our day-to-day reasoning. It is with precisely these kinds of misunderstandings that this book is concerned.

Science in the courtroom

The idea that a good scientific theory is one that could be wrong is a powerful and compelling thesis. It is a view that you will find advanced by those journalists and politicians given to the more refined pleasures of abstract philosophical reflection. It is certainly a view that you will find with predictable regularity amongst undergraduates if you have ever had the mixed pleasures of teaching an introductory course on the philosophy of science. It is even a view that you will find widely advocated by scientists themselves—from those working at the cutting edge of experimental research, sleeves rolled up, a cigarette hanging loosely from the corner of their mouth as they tinker away under the hood of the fundamental structure of reality. It is a view therefore that has unsurprisingly exerted an enormous influence over the public understanding of science. It is a view that has even formed the basis for a number of high-profile legal decisions in the U.S. concerning what can and cannot be taught as part of a high-school science curriculum.

In 1981, the State of Arkansas passed Act 590, otherwise known as the Balanced Treatment for Creation-Science and Evolution-Science

Act. While its language was a little vague, and its presentation somewhat inept, its overall message was however pretty clear: any public school in the State of Arkansas that taught students how biological complexity gradually evolved from simpler origins must *also* teach how biological complexity arose suddenly in an act of spontaneous creation. The idea was that students should be made aware of both alternatives so that they could be allowed to draw their own conclusions. Shortly after Act 590 was passed, a lawsuit was filed against the Arkansas Board of Education by a coalition of religious leaders from the local community who all professed a considerably less literal interpretation of the Bible.[2] The plaintiffs were headed by the Reverend William McLean, a United Methodist minister, and included representatives from the Episcopal, Roman Catholic, Presbyterian, and African Methodist Episcopal Church, as well as members of the Jewish community and a number of concerned teachers and parents. They argued that Act 590 was a violation of their civil rights—in addition to its well-known protection of free-speech, the First Amendment of the U.S. Constitution also explicitly forbids any governmental action pertaining to either the repression *or* establishment of a religious belief, and it was claimed that the teaching of spontaneous creation in a public school constituted a clear endorsement of Christian Fundamentalism by the State of Arkansas. After a two-month trial, expert witness testimony and a great deal of publicity, their case was eventually upheld by Judge William R. Overton.

McLean v. Arkansas was the first time that creationism had been explicitly challenged in the U.S. court system, and marked a significant turning point in a long-running and rather ignominious conflict between science and religion in the United States. The opening salvo was fired in 1925, when John Scopes had been prosecuted by the State of Tennessee for teaching human evolution in his classroom. The so-called Scopes Monkey Trial was a media frenzy, with famed politician and orator William Jennings Bryan leading the prosecution, and celebrity attorney Clarence Darrow speaking for the defense. Scopes was convicted and fined (although the decision was later reversed on a technicality). Similar skirmishes followed, although with the creationists slowly losing ground, until in 1968 the U.S. Supreme Court finally overturned existing legislation prohibiting the teaching of evolution. At this point creationists suddenly found themselves very much on the defensive, and began instead seeking "balanced treatment" for their views alongside the increasingly mainstream acceptance of the principles of natural selection. Act 590 was one of many such attempts to keep creationism on the school curriculum: a new era of guerrilla conflict that eschewed direct confrontation with the scientific orthodoxy, but rather attempted to undermine it from within. It was argued that if the scientific method is indeed committed to the spirit of open-minded and critical inquiry—and not just the promulgation of its own cherished opinions—then it should welcome the discussion of alternative points of view. In a rather ingenious intellectual maneuver, creationists argued that it is science itself that demands the equal consideration of creationism on the school curriculum.

It should be immediately noted however that the scientific credentials of creationism were in fact completely irrelevant to the issue at hand. Whatever other virtues it may or may not possess, creationism is first and foremost a religious doctrine. In the case of McLean v. Arkansas, it was pointed out that Act 590 did not simply postulate that biological complexity arose as an act of spontaneous creation, but in fact offered a detailed account of the origins of life in remarkable conformity to the Book of Genesis, right down to an explanation of the Earth's geology in terms of a catastrophic worldwide flood. It was also pointed out that while spontaneous creation does offer an alternative explanation for biological complexity, it is certainly not the *only* alternative explanation. Yet Act 590 included no provision for Buddhist cosmologies, the endless cycle of Hindu reincarnation, or any of the various Native American creation myths. Nor did it seek a "balanced treatment" for the view that living organisms first arrived on the back of an asteroid, left smeared across the surface of the Earth after some kind of interplanetary fender-bender, or that little green aliens have been directing human development from afar as part of some nefarious scheme—no doubt in cahoots with the U.S. Government and other unsavory corporations. In short, the idea that Act 590 was only seeking to ensure a balanced treatment of all of the available options was exposed as simply disingenuous: it was in fact clearly designed to promote one particular set of religious beliefs to the exclusion of all others, and was correctly ruled to violate the

First Amendment's clearly mandated separation between Church and State.

Nevertheless, the underlying idea that it was simply *good scientific practice* to give careful consideration of creationism proved to be a provocation too far. And in an act of hubris that evolutionists would later come to regret, a legal definition of legitimate scientific inquiry was duly thrashed out in the Arkansas District Court. Expert testimony was offered from a number of illustrious sources, including Francisco Ayala, an evolutionary biologist from the University of California at Irvine, G. Brent Dalrymple, a geologist with the U.S. Geological Survey, Harold Morowitz, a biophysicist from Yale, Stephen Jay Gould, evolutionary biologist and public intellectual based at Harvard, and Michael Ruse, a philosopher of science from the University of Guelph in Ontario. In his summary, Judge Overton ruled that:

> While anybody is free to approach a scientific inquiry in any fashion they choose, they cannot properly describe the methodology used as scientific, if they start with a conclusion and refuse to change it regardless of the evidence developed during the course of the investigation.[3]

In other words, the judge ruled that a good scientific theory is one that can be shown to be false—a good scientific theory is a falsifiable scientific theory. It makes precise and substantive claims about the nature of the world that could turn out to be false, and when they do, the theory is amended, refined, or simply abandoned altogether in favor of an alternative account that is better supported by the evidence. None of these characteristics were found to apply to creationism. Therefore, in the opinion of Judge Overton and the Arkansas District Court, it is not science.

The issue seemed to be pretty clear cut. We begin with a rigorous definition of science, one that is both intuitive and widely-held amongst both the general public and the scientific community. The definition is applied in order to assess the scientific credentials of creationism, and it fails miserably. The idea that biological complexity arose on Earth from an act of spontaneous creation is just not the sort of claim entertained by legitimate scientific inquiry. In 1987, the decision of McLean v. Arkansas was upheld by the U.S. Supreme Court in response to a similar challenge in the State of Louisiana. Falsifiability was on the books. *Yet somehow it didn't seem to make the slightest bit of difference.* Creationism did not go away, but in fact came back stronger than ever, and in 2004 the Dover Area School Board in the State of Pennsylvania passed an act requiring the "balanced treatment" of evolution alongside something called "intelligent design theory"—a slick repackaging of creationism, superficially divorced from any overt religious overtones, and carefully crafted to satisfy the existing legal understanding of legitimate scientific inquiry.

So what exactly went wrong? How did our intuitive understanding of scientific practice as open-minded and critical inquiry, hammered out by experts and rigorously applied by a court of law, fail so spectacularly in quashing such a paradigmatic example of pseudo-scientific dogma? One possibility of course is that creationism is in fact far more scientific than we initially supposed, and that its continued survival is testimony to the fact that deep down it really does satisfy our intuitive convictions of what makes a good scientific theory. Another possibility is just that creationists are very good at rebranding, and that they are particularly adept at packaging their ideas in such a way as to slip under the legal radar. Quite probably, it is some combination of the two. But I want to propose a third, and far more interesting possibility—namely that our intuitive understanding of scientific practice is fundamentally flawed, and that good science has absolutely nothing to do with falsifiability whatsoever.

Science and pseudoscience

It is no great surprise that the notion of falsifiability should figure so prominently in the case of McLean v. Arkansas. In his own attempt to explore and articulate the idea, Karl Popper—whose work was indeed explicitly cited in expert testimony throughout the case of McLean v. Arkansas—argued explicitly that falsifiability offered a way for us to distinguish between the genuine scientific practice that we should admire, and the pseudoscientific dogma that we should reject. It was a criterion specifically designed to weed out those disreputable claims and worthless theories that could only masquerade as legitimate empirical inquiry. In other words, it was a tool custom made for the job at hand.

The background to Karl Popper's intellectual development is instructive at this point. Growing up in Vienna during the interwar period, he was to experience some of the worst excesses of totalitarian government, as well as some of the most groundbreaking successes of modern science. These events were to exert a profound influence upon him. Although initially attracted to Marxism, Popper quickly grew disillusioned with what he saw as the false promises of yet another self-serving ideology, and watched in horror as the forthcoming socialist utopia quickly degenerated into its own brand of sectarian killings. However, at about the same time as he was coming to critically reassess his political commitments, Popper was also struck by the spectacular confirmation of Einstein's general theory of relativity. During a solar eclipse in May 1919, Arthur Eddington successfully observed the deflection of light in the presence of a gravitational field. It was an extraordinary result that had required months of careful preparation, and which had taken the intrepid astronomical team to a small island off the western coast of Africa in order to make their observations. It was also an extraordinary risk—the bending of light was such an absurd and unprecedented prediction that many in the scientific community simply took it as evidence of the implausibility of Einstein's entire approach. The successful outcome made front-page news around the world. Eddington had put the general theory of relativity to the most severe test, and it had passed with flying colors.

For Popper, the bold conjectures and critical assessment of scientific practice stood in stark contrast to the narrow-minded indoctrination of Marxism. The distinction was to motivate two closely-related questions which would concern Popper for the rest of his career: How exactly does the scientific method work? And how can we distinguish between genuine scientific theories like Einstein's theory of relativity, and those pseudoscientific dogmas like the Marxist theory of history that only masquerade as legitimate empirical inquiry? The central idea of Popper's work is that these two questions are in fact *the same question*, and can be answered in exactly the same way. The distinctive feature of scientific practice responsible for its success—and which by contrast is noticeably absent from all forms of pseudoscientific indoctrination—is an emphasis upon *critical testing*. So for example, the importance of Eddington's experiment lies in the fact that if the predicted deflections had not been observed, then the general theory of relativity would have been conclusively refuted. It was the riskiness of the experiment that made the positive result so important. By contrast, Popper argued that the problem with something like Marxism is that it does not appear to be vulnerable to the same sort of empirical challenge. Whether the eventual revolution of the proletariat takes place in highly industrialized countries, as Marx originally predicted, or whether it eventually takes place in an economically backwards country, as realized by the Russian Revolution, both eventualities can be explained just as well in terms of a sufficiently vague notion of a "class struggle" (note the inverted commas again). When the workers do what they are supposed to do, it is because Marxist theory is correct. When the workers do not do what they are supposed to do, it is because they are suffering from "false consciousness" and have been corrupted by the evils of capitalism. In Popper's view, since Marxism takes no risk in making its predictions, it cannot take any credit when those predictions are confirmed.

Another favorite example that Popper liked to discuss was the field of psychoanalysis—a paradigmatic example of junk science if ever there was one—and in particular the theories of Sigmund Freud and Alfred Adler with which he became familiar in Vienna. In both cases, Popper argued, there is no possible human behavior that cannot be accommodated by these theories. If an individual doesn't act in accordance with his diagnosis, this is not because he doesn't have such a condition, but only that he is "in denial". In his later work, Popper makes this idea explicit. He writes:

> I may illustrate this by two very different examples of human behavior: that of a man who pushes a child into the water with the intention of drowning it; and that of a man who sacrifices his life in an attempt to save the child. Each of these two cases can be explained with equal ease in Freudian and in Adlerian terms. According to Freud the first man suffered from repression (say,

of some component of his Oedipus complex), while the second man had achieved sublimation. According to Adler the first man suffered from feelings of inferiority (producing perhaps the need to prove to himself that he dared to commit some crime), and so did the second man (whose need was to prove to himself that he dared to rescue the child). I could not think of any human behavior which could not be interpreted in terms of either theory. It was precisely this fact—that they always fitted, that they were always confirmed—which in the eyes of their admirers constituted the strongest argument in favor of these theories. It began to dawn on me that this apparent strength was in fact their weakness.[4]

Popper's insistence upon critical testing seems therefore both to capture those aspects of open-minded scientific practice that we admire, and by contrast to highlight those features of pseudoscientific quackery that we despise. Moreover, it is philosophically very satisfying. Popper's proposal is both simple and precise. It is hardly surprising therefore that such a framework should find itself taking pride of place in the legal dispute over creationism.

But there is a problem. For while Popper's views have been extremely influential on the public understanding of science, it is probably fair to say that he has not been taken very seriously by contemporary philosophers. This is due in part to the inevitable fashions of academic life, and Popper's own unrivalled ability to lose friends and alienate his colleagues. For a philosopher who advocated the values of open-minded and critical discussion above all else, Popper was notoriously hostile to any criticisms of his own views, and demanded absolute orthodoxy amongst his own students. But there are also genuine difficulties with Popper's account of the scientific method. The emphasis upon testability and falsification is an extremely simple and powerful idea, and one which Popper was able to apply to a great deal of different issues. But the realities of scientific practice are rarely so straightforward, and Popper devoted much of his later work to dealing with all of the exceptions, anomalies, and counterexamples that continuously plague this otherwise elegant account. As with all simple ideas, the devil is in the details.

Finding Neptune

The first thing that we have to note is that the account of falsification presented above is of course something of a simplification. We rarely—if ever—manage to subject our scientific theories to anything like a conclusive case of falsification. This is because we never test our scientific theories in *isolation*, but rather as part of a larger theoretical network. Many of the predictions made by our scientific theories presuppose the predictions made by other scientific theories, or require further scientific investigation in order to assess properly. For example, suppose that we are attempting to test a particular theory of planetary motion, and that we predict that a specific heavenly body is going to be found at a specific region of space at a specific moment of time. One important assumption that we are going to have to make in a case like this is that there are no other complicating factors to take into consideration, such as rogue asteroids exerting their own gravitational influence, or cataclysmic events on the planet's surface sufficient to derail its otherwise regular orbit. We may of course have good reasons to suppose that none of these eventualities are in fact the case, but the point is that we frequently have to rely upon scientific theories *other than the one under test* in order to make our predictions. So in this example, our prediction about the future location of a particular planet will in actual fact constitute a test for both our theories of planetary motion and exogeology, since errors in either one of these two theories will yield inaccurate predictions.

Moreover, in order for us to actually test this prediction, we will have to do more than simply look up at the night sky. The planet in question may well be too distant for us to observe without sophisticated equipment, and at the very least, we will need to take accurate measurements if we are to perform a rigorous test of our scientific theories. So in order to test our prediction, we will probably use a high-powered telescope, or some other method for detecting electromagnetic radiation, pieces of equipment that are themselves constructed on the basis of a whole background of often quite complex scientific theory. Again of course, we may have countless good reasons to suppose that our scientific equipment is reliable;

but again, we also have to note that one way in which we may fail to observe the predicted location of our orbiting planet is that our instruments are incorrectly calibrated, faulty, or damaged. So it turns out that our prediction about the future location of a particular planet will in fact constitute a test of our theory of planetary motion *and* our theory of exogeology, *in conjunction with* certain assumptions about the initial conditions of the planet in question and the absence of any other intervening factors, *taken together with* our theories of optics, the propagation of electromagnetic waves, the physiology of the human eye, and so on and so forth.

The issue here is not merely practical—of course science is a complex and messy business, involving a great number of moving parts. The problem is how exactly this greater complexity impacts upon the testing and potential falsification of an individual scientific theory. We predict the location of an orbiting planet, and fail to observe it at the proper time. Clearly something has gone wrong. But where exactly does the fault lie? Is our theory of planetary motion incorrect? Or was there some external factor we failed to take into account? Perhaps our telescope was malfunctioning? Maybe the graduate student we left alone all night in the observatory fell asleep or got distracted watching his favorite Jean-Claude Van Damme movie? And unfortunately, the failed prediction alone cannot tell us which part of our vast theoretical network is to blame, and thus we can never tell for sure if we have falsified our theory of planetary motion, rather than any other of the numerous scientific theories that we relied upon in generating our original prediction. In short, the falsification of a scientific theory always involves far more than just the logical relationship between theory and prediction. Given the complex interrelationships between our scientific theories, and the additional background assumptions that must be made in order to actually make any predictions, there is no simple algorithm for determining which part of our theoretical network has been shown to be false. Rather, there is always a *decision* that has to be made about how we are to interpret the results of any such experiment.

But once we acknowledge the fact that scientists will have to choose how exactly they interpret the results of their experiments, our initially simple picture begins to fall apart. We have to ask how these decisions are made, and whether or not they bring an element of personal preference or bias into the process of science. At the very least, we have certainly moved away from the ruthlessly logical framework that Popper proposes. Moreover, such an approach has in fact yielded some astonishing successes. One of the best examples of this concerns the discovery of Neptune. In the early nineteenth century it was noticed that Newtonian Mechanics failed to accurately predict the orbit of the planet Uranus, which was believed at the time to be the outermost planet of the Solar System. A number of possibilities were proposed to explain this fact, from inaccurate telescopes to incompetent astronomers. At no point however did anyone suggest that Newtonian Mechanics had been falsified and should therefore be rejected. The theory was just too well entrenched for this to be a serious consideration. Rather, it was eventually proposed that there must exist a mysterious extra planet lying beyond Uranus, whose gravitational pull was sufficient to cause the observed perturbations. Such a proposal was clearly ad-hoc, since the only reason anyone had for supposing that this extra planet existed was the fact that Newtonian Mechanics was unable to accurately predict the orbit of Uranus. Nevertheless, a desperate race began to locate this mysterious planet, and in September 1846 a bitter priority dispute duly began between British and French mathematicians as to who had calculated its position first (the French eventually won).

Falsifying a scientific theory turns out therefore to be much more complicated than we originally thought. Since we only ever test our scientific theories in groups, it can be difficult to work out *which* scientific theory is being tested by any particular experiment. But more importantly, even when we do seem to have a straightforward case of falsification, it is not always good scientific practice to act upon it. Sometimes the best scientific course of action is to ignore the falsification, and try to find some way to preserve our scientific theory in the face of contrary evidence. And this leaves us in something of a difficult predicament. If good scientific practice can sometimes involve ignoring a falsifying experiment, or amending our theories in order to accommodate evidence to the contrary, just how exactly does this differ from the sort of conspiracy theories with which we began? I claim that the moon landings were faked in 1969, presumably in order to improve Nixon's chances of reelection. When you point out that the lunar landing sites can be observed

FIGURE 1.1 *A British astronomer discovers Neptune ... in the calculations of his French rival. While obviously intended to poke fun at the unsuccessful Brits, the cartoon nevertheless unintentionally makes an important point about the entire incident. Neptune was not discovered by observing the night sky, but through careful mathematical calculation. The British astronomer has the right idea.*

through the Hubble telescope, I respond that the photographs have been doctored in order to preserve the hoax. It certainly seems unlikely—but for astronomers working in the nineteenth century, it also seemed pretty unlikely that there could be an entirely new and previously undiscovered planet lying just beyond the orbit of Uranus.

In the end of course it all comes down to a matter of degree. If our nineteenth-century astronomers had consistently failed to observe Neptune, they might have made some other adjustments in their theory. One can always blame the lazy graduate student or an insufficiently accurate telescope. Perhaps there was *another* undiscovered planet lying just beyond the orbit of Neptune, whose gravitational pull was sufficient to divert it from its expected location. If these adjustments also failed to deliver the correct result, perhaps one more planet could be added to the mix. At a certain point, however, it seems reasonable to suppose that the astronomers would finally admit defeat, and abandon Newtonian Mechanics altogether. By contrast, the committed conspiracy theorist will go on spinning ever more complex webs of deceit forever. There is after all no limit to Nixon's evil genius. But this has nothing to do with the theories themselves, but rather the various individuals who endorse them. There is nothing about Newtonian Mechanics that stipulates that you can only posit one or two extra planets before rejecting the theory altogether—if necessary, you can keep on adjusting the theory for as long as you like. When we get down to the technical details of it, Newtonian Mechanics turns out to be just as unfalsifiable as any conspiracy theory.

Science and pseudoscience revisited

We began this chapter with the idea that a good scientific theory is one that is falsifiable. On closer examination, it turned out that any scientific theory can be preserved in the face of any evidence to the contrary—provided of course that we are willing to make the necessary adjustments elsewhere in our beliefs. In that sense then, there is no difference at all between the claim that force is equal to mass times acceleration, and the claim that Nixon faked the moon landings. The important difference is that while any scientific theory can be maintained indefinitely against refutation, there comes a point at which the scientific community will eventually abandon the idea. But this is a fact about scientists and the attitudes they take towards their theories. It has nothing to do with the nature of the theories themselves.

This tendency to conflate the falsifiability of a scientific theory with the open-mindedness of the scientists who believe them is unfortunately extremely common. It is a mistake clearly made in the case of McLean v. Arkansas. As you will recall in his summary, Judge Overton ruled that anyone who refused to revise their conclusion in the face of recalcitrant evidence could not describe their methodology as scientific. But that does not actually tell us anything at all about the scientific theory in question. It may well be the case that creationists are particularly stubborn when it comes to the revision of their cherished world view. But that would be a fault of *creationists*, not creationism. The important question is whether or not the view that life began in an act of spontaneous creation some 6,000 years ago is open to empirical refutation—and not the extent to which its advocates have actually subjected that view to the requisite degree of testing. After all, if someone working on the cutting edge of quantum mechanics showed a pronounced reluctance to modify their pet hypothesis in the face of conflicting experiment, we might well deplore their lack of open-mindedness, but we would not thereby reject quantum mechanics itself as an unscientific enterprise.

The point is an important one, since once we clarify this deliberate conflation between theories and theorists, we can see that creationism actually *satisfies* the characterization of science reached in the case of McLean v. Arkansas. Let us grant for the sake of argument that spontaneous creation offers us an unequivocally flawed account of the origins of the life, one that is unable to adequately accommodate the fossil record, or to satisfactorily explain the full-range of biological complexity we encounter in the world today. Nevertheless, for all its many faults, none of this shows that creation-science is unfalsifiable. On the contrary, creation-science makes a number of specific, highly testable claims about the age of the earth, the geological consequences of a worldwide flood, and the relatively limited degree of variability to be found amongst the animal kingdom (at least as compared to that expected by the process of random mutation and natural selection). The problem with creation-science is not that it appears to be unfalsifiable, but rather that it has been repeatedly *falsified*. The predictions it makes have just not been borne out. But that doesn't mean that creation-science is pseudoscientific nonsense. Precisely the opposite, in fact. If our sole criterion for being genuinely scientific is that a theory be falsifiable, then it seems that we have to conclude that anything which has in fact already been falsified is about as genuinely scientific as it can get. Despite a wide-spread conviction to the contrary, the reasoning in McLean v. Arkansas—if rigorously applied—would actually legitimize the scientific status of creationism.

All of which is a somewhat disturbing turn of events. The situation however is further compounded by the fact that Popper himself maintained throughout his career that it was the *theory of evolution* that was in fact unfalsifiable. He frequently complained that the theory was tautological, and that any evidence whatsoever could be shown to be compatible with it. As Popper put it:

> To say that a species now living is adapted to its environment is, in fact, almost tautological. Indeed we use the term "adaptation" and "selection" in such a way that we can say that, if the species were not adapted, it would have been eliminated by natural selection.[5]

For example, we might suppose that the theory of evolution predicts the ever increasing variety of organisms through the continuous process of mutation and environmental pressure, such that we might reasonably expect to see more and more species as time goes on. Or we might suppose that the theory of evolution predicts that any complexity that arises in an organism will be such as to provide an advantage in the organism's environment. But neither are, in fact, the case. If the circumstances in question are especially harsh—a desperate struggle over a very limited range of food and shelter, for instance—it may well be the case that only a handful of different survival strategies will be feasible, and that therefore there will not be a great deal of variation between surviving species. On the other hand, if the circumstances in question are particularly favorable—abundant resources and limited predators—then a vast range of useless adaptations may flourish in the absence of any genuine pressure against them. In short, it seems that almost any eventuality is compatible with the mechanism of natural selection, given the right sort of circumstances.

Later in his career, Popper did come to qualify his position. He acknowledged that many of the theoretical underpinnings of the theory of evolution—for example, theories describing the mutation, recombination, and inheritance of genetic material—are legitimate scientific theories capable of rigorous testing and falsification. Nevertheless, Popper continued to maintain that the broader theoretical framework, as articulated in terms of the principle of adaptation through natural selection, was an untestable hypothesis and therefore could not be considered as a genuinely scientific explanation for biological complexity.

Back to the courtroom

In 2006, science and religion met again in the Pennsylvania District Court. This time the plaintiffs were a group of concerned parents led by Tammy Kitzmiller, and the issue concerned the proposal by the Dover Areas School Board to offer a "balanced treatment" of evolutionary biology and intelligent design theory—which they assured everyone was definitely *not* a cynical and superficial rebranding of creationism, but something completely different altogether. The ploy fooled no one, and Judge John E. Jones III took little time to uphold the complaint, ruling that intelligent design theory was still a straightforwardly religious doctrine and that its inclusion on the high-school curriculum clearly violated the First Amendment's separation between Church and State. Yet just as in the case of McLean v. Arkansas, the scientific credentials of intelligent design theory were to feature prominently in the decision. In the case of Kitzmiller v. Dover, however, a very different definition of science was proposed.

Expert testimony was again gathered from the great and the good, in this instance by Kenneth R. Miller, a biologist from Brown University, John Haught, a theologian at Georgetown University, Robert T. Pennock, a philosopher of science at Michigan State University, and Barbara Forrest, a philosopher at Southeastern Louisiana University. The notion of falsifiability was dropped altogether in favor of a broader description of scientific practice. It was proposed that in general, good scientific practice is committed to a principle of *methodological naturalism*—the idea that the world is governed by natural processes, and that a good scientific theory cannot appeal to miracles or other supernatural forces as part of its explanation for how that world works. In his ruling, Judge Jones summed this up by saying:

> After a searching review of the record and applicable case-law, we find that while intelligent design arguments may be true, a proposition on which the Court takes no position, intelligent design theory is not science ... Expert testimony reveals that since the scientific revolution of the sixteenth and seventeenth centuries,

science has been limited to the search for natural causes to explain natural phenomena.[6]

Obviously enough, intelligent design theory violates the principle of methodological naturalism by invoking a supernatural force responsible for an act of spontaneous creation. Therefore, in the opinion of Judge Jones and the Pennsylvania District Court, it is not science.

It should be clear enough from the foregoing discussion why the notion of falsifiability was rejected as providing the demarcation between science and pseudoscience. Nevertheless, it is fair to say that the reasoning offered in Kitzmiller v. Dover would have benefited considerably if it had continued to bear such considerations in mind. To argue that a good scientific theory is one that only appeals to natural causes in order to explain natural phenomena may seem plausible at first inspection, until we realize that it is only on the basis of our scientific theories that we have any grasp at all of the notion of a natural cause. Let us put the point another way. We would all agree that it would be illegitimate for a scientific theory to invoke the existence of ghosts and goblins in order to explain why the Earth orbits the Sun, or why magnets attract iron filings, or anything else for that matter. But the reason why we reject the background machinations of ghosts and goblins is because our best scientific theories tell us that they do not exist. It is science itself that determines what counts as a "natural cause" or a "natural phenomenon." To be "supernatural" therefore means nothing more than to lie outside the realm of scientific investigation. But in that case, to say that a good scientific theory is one that only appeals to natural causes is to say that a good scientific theory is one that only appeals to the things that science talks about. The notion of methodological naturalism is thus *completely empty*—it is unfalsifiable in the worst possible way, amounting to little more than the claim that good science must be appropriately scientific.

And this is a truly alarming state of affairs. The legal conflict between science and religion began with a definition of science intended to discredit creationism, but which actually cast doubts on the scientific credentials of evolution. However, since no one involved in McLean v. Arkansas seemed to understand anything they were saying, the faulty criterion was duly misapplied to achieve the required result. When creationism repackaged itself, an amended definition of science was proposed that was not only utterly meaningless, but which turned out to be guilty of the very same intellectual dishonesty the first definition was supposed to remedy. If this is the sort of reasoning about science that constitutes the basis of wide-spread legal decisions, then we might well wonder what possible chance we have of distinguishing between "genuine scientific inquiry" and "pseudoscientific dogma" in the first place.

But there is also a deeper problem here. It is one thing to want to remove creationism or intelligent design theory from the classroom. There are plenty of good reasons to prefer the theory of gradual evolution as the explanation for biological complexity, and as a framework for its continued investigation. To take one very simple example, it is not clear what kind of *research projects* would be suggested by creationism, or what kinds of *experiments* we might be encouraged to perform. With evolution at least, we have a potential mechanism that explains how things operate, and which we can attempt to manipulate in order to investigate the world and to try and improve it. Tinkering around with DNA and the like can help us cure diseases. But if everything in the world has been put that way by an all-powerful deity, then it is not clear how we should go about trying to improve our lot (or even if we should). Notice that none of this depends upon whether or not creationism or evolution are actually true. It is a practical argument for favoring evolution because it provides a more useful framework for future investigations. It is an argument for teaching evolution in the classroom that even the most committed creationist could accept.

But either way, this should not be a matter for *legislation*. It should be a matter for the marketplace of ideas, for discussion and debate. And again, this is not an argument based upon any political attachment to small government or minimal state intervention. It is a practical argument that acknowledges that the best way to eliminate a bad idea is to accept it on its own terms and to subject it to critical scrutiny—for once you drive an idea underground, it will only grow in popularity. Yet by seeking a legal definition of science by which alternative points of view can be legitimately dismissed, this is precisely what we are doing. It is therefore of no great surprise then that creationism bounded back from the ruling in McLean v. Arkansas in the way that it did, nor that the modified definition offered in Kitzmiller v. Dover was so fantastically blatant in its purpose. It was a position that argued that by definition, belief in God is not scientific, therefore your opinions do not count. The entire debate was therefore a piece of pure politics, both on the part of the proponents of creationism, and on the part of those who sought to exclude it altogether from the public sphere.

Nevertheless, I believe that two important lessons can be drawn from this fiasco. They are not particularly inspiring lessons, but they remain important nevertheless. The first is simply that the scientific method is both complex, and extremely poorly understood. The various definitions legislated by the courts—that a scientific theory is a falsifiable theory, and that a scientific theory is a naturalistic theory—are truly dreadful. They are definitions that fail to accommodate many paradigmatic instances of good scientific practice, frequently legitimize paradigmatic examples of poor scientific practice, and in the case of methodological naturalism at least, are almost entirely empty of anything resembling intellectual content. Moreover, these were definitions suggested by some of the practicing scientists themselves. We should therefore be extremely wary of anyone who claims to have identified the essential elements of the scientific method, and subject these claims to close scrutiny.

The second lesson is that it is often very easy to confuse what one believes to be good scientific practice with what one believes to be a politically desirable outcome. Science plays an enormous role in our lives, and is increasingly appealed to as an arbiter in our social and political interactions. There is after all no better way to silence an opponent that to denounce their opinions as "non-scientific." This problem, however, is exacerbate by our first problem, and the fact that no one seems to know what exactly it means to be "scientific" in the first place. It is usually taken to be a kind of shorthand for being rational or offering reasonable arguments—but in that case, it becomes even easier to conflate what one believes to be good scientific practice with what one believes to be a politically desirable outcome, since no one holds a political view without what they believe to be rational reasons. In this way, the political cart often leads the scientific horse, often with unfortunate consequences.

While the details of his approach leave a lot to be desired, Popper did make a good point when he compared the open-minded spirit of scientific inquiry with the principles of democratic government. It is a lesson that seems to have been lost.

2
A matter of trial and error

I want to tell you another story about the scientific method. Stop me if you've heard it before.

It is a bright, sunny morning in the summer of 1591. In the sleepy university town of Pisa, an enormous crowd has gathered outside the cathedral, under the shadow of its famously lopsided bell-tower. The atmosphere is lively and festive, and people bustle around expectantly, talking excitedly with their neighbors while unsupervised children dart about between their legs. Many arrived early to ensure the best vantage points; others lurk near the back, drinking, laughing, and placing bets. Suddenly a hush descends upon the crowd. A procession appears from the nearby university—a cacophonous gaggle of scruffy academics and sour-faced clerics, trailed by their attendant army of acolytes and graduate students. They reach the foot of the tower. There is a heated discussion amongst some of the most prominent members of the academy, which eventually breaks down into a full-blown argument. Unpleasant words are uttered and colorful gestures exchanged. Pushing and shoving breaks out across the crowded piazza, and groups of students begin to coalesce into furiously antagonistic factions. Eventually a compromise is reached and calm is restored. All eyes now turn back to the tower. A single individual separates from the group, and struggling under the weight of two vastly mismatched cannonballs, begins to climb the 296 stairs to the top.

The man is Galileo Galilei, who at the age of twenty-five has already been appointed professor of mathematics at the university, and is quickly establishing himself a reputation as the *enfant terrible* of the scientific world. According to the authority of Aristotle, everything in nature has its natural resting place, and just as fire burns upwards in an effort to rejoin the heavens, so too does matter possess an innate desire to return to the center of the Earth. The greater the quantity of matter, the greater the desire, and thus it stands to reason that a heavier object will fall faster than a lighter object whenever the two are dropped from a great height. This is the undisputed scientific consensus of the day, and if you were to query any of the many experts assembled on that sunny morning in Tuscany, they would be happy to show you the passages in their well-thumbed copies of Aristotle's *Physics* where he presents the principles of motion that have dominated European thought for nearly 2,000 years. But Galileo disagrees. He claims that different objects of different weights will in fact fall at the same speed. The suggestion is unthinkable—even the Church endorses the truth of Aristotle! Eventually Galileo reappears at the top of the iconic tower. He places the two unevenly weighted cannonballs side-by-side on the parapet and gently nudges them over the edge. The crowd holds its breath. No one dares to blink. And then, with an almighty thud, the two cannonballs strike the ground in perfect accord. A great roar comes up from the crowd. The students are ecstatic. The elderly academics shake their heads in disbelief. Some of the clergy can be seen crossing themselves in horror. Galileo is promptly arrested by the Inquisition and thrown into a dungeon. He is tortured and eventually recants his heresy. But it makes no difference. The truth is out, and modern science is born.

Thus goes one of the great foundation myths of modern science, one which we are all taught at school, and one which we all teach our own children in turn. It is a good story, but most of its appeal lies in the fact that it manages to encapsulate everything that we believe to be at the very heart of the scientific method—the challenging of received wisdom, the importance of experiment, and the eventual triumph of careful and meticulous observation over the blinkered indoctrination of authority. It offers a detailed microcosm of the seemingly perennial conflict between science and religion, and establishes the indelible blueprint for our popular image of the isolated scientific genius struggling against the forces of the conservative status quo. It is in short one of the most important experiments ever to be performed in the history of science.

The problem is that it never happened.

Our suspicions should perhaps have been raised by the fact that Galileo himself never actually mentions any such experiment at all throughout his voluminous notes and correspondences of the period. Nor do any of his contemporaries. For such an apparently momentous and epoch-making experiment—one that was also by all accounts such a celebrated public spectacle—it is somewhat remarkable that absolutely no one at the time seems to have noticed that it ever even occurred. There is no record of any controversy during Galileo's time at Pisa, and it would take another forty years before he finally came to blows with the ecclesiastical authorities. It is also far from clear whether Galileo had even fully formulated his opposition to the Aristotelean theory of motion during his time at Pisa, let alone designed and performed the dramatic refutation with which he is credited. The earliest written reference to the experiment does not in fact occur until 1654—that is to say, over sixty years later—in what can only be described as an excessively heroic biography written several years after Galileo's death by his friend and pupil Vincenzo Viviani. The whole incident seems to be nothing more than a piece of literary romanticism which has somehow taken on a life of its own to become one of the most successful examples of scientific propaganda ever created.

But the situation is in fact even more alarming than it appears at first sight. The problem is compounded by the fact that while Galileo never actually performed his most famous experiment, some of his contemporaries *did* drop cannonballs from the top of the Leaning Tower of Pisa—and got precisely the opposite result! In 1612, Giorgio Coresio, another professor at the University of Pisa, conducted this exact experiment, not in support of Galileo's conjecture but rather in an attempt to confirm the traditional belief that heavier bodies fall quicker. Complaining about the work of one of his rivals, Jacopo Mazzoni, a professor of philology and literary criticism at Pisa who nevertheless also found the time to criticize Aristotle's theory of motion, Coresio writes that he:

> commits anew two new errors of no slight importance. First, he denies a matter of experiment, that with one and same material, the whole moves more swiftly than the part. Herein

his mistake arose because—perhaps—he made his experiment from his window, and because the window was low all his heavy substances went down evenly. But *we* did it from the top of the cathedral tower in Pisa, actually testing the statement of Aristotle that the whole of the same material in a figure proportional to the part descends more quickly than the part. The place, in truth, was very suitable, since if there were wind, it could by its impulse alter the result; but in that place there could be no such danger. And thus was confirmed the statement of Aristotle, in the first book of *De Caelo*, that the larger body of the same material moves more swiftly than the smaller, and in proportion as the weight increases so does the velocity.[1]

Of even greater interest is the fact that a few years later in 1641, one of Galileo's own pupils—Vincenzo Renieri—also dropped cannonballs off the top of the Leaning Tower of Pisa, and wrote to his former teacher for help in interpreting his results:

We have had occasion here to make experiment of two weights falling from a height, of diverse material, namely one of wood and one of lead, but of the same size; because a certain Jesuit writes that they descend in the same time, and with equal velocity reach the earth … But finally we have found the fact in the contrary, because from the summit of the Campanile of the Cathedral, between the ball of lead and the ball of wood there occur at least three cubits of difference. Experiments also were made with two balls of lead, one of a size equal to a cannon-ball and the other to a musket-ball, and there was observed between the biggest and the smallest, from the height of the Campanile, to be a good palm's difference by which the biggest preceded the smallest.[2]

It is noticeable that Renieri forgets to mention the fact that Galileo himself had supposedly conducted the very same experiment fifty years earlier. It is also noticeable that just like Coresio before him, Renieri also failed to get the expected result. For an experiment that supposedly triumphs the virtues of disinterested observation and trusting the evidence of our senses, there seems to be a lot more going on in this simple story than immediately meets the eye.

Seeing is believing

The truth is that there is absolutely no evidence whatsoever that Galileo ever conducted his most celebrated of experiments. He certainly never mentions it himself—and neither his closest friends, nor his most embittered opponents, seemed to be aware that it had ever taken place. But leave all of this to one side. There is in fact a much bigger difficulty with the story, and with our whole foundation myth of modern science. The experiment described in Galileo's biography is impossible. It simply *could not* have occurred in the way in which it was described. For while Galileo was certainly correct that different bodies of different weight will nevertheless experience exactly the same rate of acceleration due to gravity, this does not necessarily mean that they will fall at the same rate, since we also now know that the effects of air resistance will be significantly different. In simple terms, the lighter the falling body, the more the impeding force of air resistance will be felt in proportion to its mass, and thus the greater the influence it will have upon its overall speed. Indeed, I invite you to conduct the same experiment yourself. Go to Pisa and throw some heavy objects off the top of the tower. Chances are that you will kill some tourists, but once the chaos and legal proceedings have passed, you will find yourself in exactly the same situation as Coresio and Renieri. The heavier cannonball *will* reach the ground first. Not by a lot of course, and certainly not to the extent suggested by some of Aristotle's more vehement supporters at the time of Galileo. But nevertheless, the heavier cannonball will land first. Ironically enough then, rather than proudly tracing their pedigree to that sunny morning in Tuscany, it is precisely our modern theories of mechanics that prove that Viviani's story is a fabrication. If there had been a crowd gathered in the Piazza del Campanile, all they would have seen was a confirmation of the prevailing Aristotelean orthodoxy.

This does not mean however that the incident cannot teach us something important about observation and experiment, and about the inner workings of the scientific method. The story was supposed to demonstrate how good scientific practice lies in conducting experiments and simply observing the results—the process of trial and error, as my old school textbooks used to put it. But a closer analysis shows us that observation is not such a straightforward process. When we drop cannonballs from a great height, our modern understanding of mechanics teaches us to observe two bodies undergoing the same rate of acceleration, but nevertheless reaching different terminal velocities due to the varying effects of air resistance. But when Giorgio Coresio dropped cannonballs from the top of the Leaning Tower of Pisa, he simply observed the truth of the traditional Aristotelian world view, since after all the two bodies did fall at different speeds. Poor Vincenzo Renieri—still struggling to make the transition between the two scientific world views—wasn't quite sure what he had observed one way or the other. And that is why if Galileo *had* released his cannonballs from the Campanile back in 1591, it would almost certainly have been perceived to be a spectacular failure. There was certainly nothing about the observations themselves that would have decided the issue. Look! The heavier cannonball landed first! For the crowds gathered in the piazza, the whole incident would have been seen to be just another vindication of Aristotle and the prevailing scientific consensus. Galileo would have been laughed off the stage. He would have been forgotten, forever consigned to the footnotes of history.[3]

The point then is not that Galileo was wrong or that our modern theories of mechanics are mistaken. It is simply that any observation requires an interpretation, and that an Aristotelian would have been just as capable of interpreting these results within his own scientific framework as any contemporary scientist would be able to today. This then is perhaps the real lesson of the story. There are any number of ways in which our scientific theories can influence the observations we make. One obvious example is the fact that we have to rely upon our scientific theories in order to help us *design* our experimental tests. Conducting an experiment is hard work, because we need to isolate the particular phenomenon that we want to study from everything else that is going on around it. We have to eliminate any extraneous factors that might interfere with the experiment, or otherwise be capable of corrupting our results. One of the reasons why both Coresio and Renieri decided to conduct their experiments from the top of the Leaning Tower of Pisa was that they thought it would allow them to control two other important variables that their rivals had not—that the two cannonballs had a sufficiently long drop to complete, and that there were no cross-currents or side-winds that could influence their descent. But this obviously requires some background theoretical knowledge. For unless you already have reasons to believe that the differences between the two cannonballs will only become apparent after a sufficiently long time in the air, you would have no more reason to perform the experiment from the top of a tower than from the top of your kitchen table.

We also need to rely upon our scientific theories to help us *discriminate* between our experimental results. Any single experiment will produce an insurmountable quantity of data, most of it completely irrelevant to the question at hand. When we drop cannonballs off the Leaning Tower of Pisa, we are interested in how long it takes for them to reach the ground. We will therefore want to pay close attention to things like the weight of the cannonballs and the time it takes for them to fall, and we will generally ignore factors like what color shirt Galileo is wearing when he throws the cannonballs over the side, or the reaction of the crowd when they finally come to rest. But all of these details are just as much a part of the experiment as any other. It is all a matter of context, and if we were interested in testing a different type of scientific theory—say, a sociological investigation into the fashion-sense of scientists, or the importance of spectacular demonstrations for the public consumption of science—we might well take a very different attitude to what is important. In short, we often have to rely upon our scientific theories to help us determine which observations are even worth the trouble of making in the first place.

Finally, there might even be a sense in which our background scientific theories can actually help to *determine* the content of our observations. It is certainly the case that different expectations can have an important effect on what we see—a fact that is exploited in any number of familiar optical illusions or magic tricks. But at a more fundamental level, it seems that any act of observation must require some kind of framework in order for that activity to make any sense. It is certainly tempting to think of observation as a purely passive activity, and that once we have isolated the object of our investigations and made the necessary provisions for all the different factors that could influence our experiment, all we have to do is open our eyes and wait for the results to arrive. But just take a moment to think about what that would really involve. Without some way of organizing and structuring that information, our sensory input would be just a barrage of colors and sounds. As the great American philosopher and father of psychology William James once put it, the world would be nothing more than "one great blooming, buzzing confusion"—and definitely not the sort of thing that could either confirm or refute a detailed scientific experiment.[4]

Thus in order to actually observe the things that are going on around us, they have to be put into context and conceptualized. In the case of our scientific experiments, this means that the things that we see have to be incorporated within some kind of theoretical framework. They have to be seen *as* something—a falling body, an accelerated mass, an impeded natural motion. It is not enough to simply drop cannonballs from the top of a high tower and stare with rapt attention at their motion, for unless these moving flashes of color and shape are interpreted as physical bodies falling towards the Earth at a greater or lesser speed, the entire event would not even count as a relevant observation. The problem then is that there are different ways of fleshing out this kind of interpretation. For Aristotle, a physical body was something that fell faster in proportion to its weight. For Galileo, a physical body was something that only moves if accelerated by a force. If they had both been present at the Campanile in 1591, they really would have seen very different things.

Reflections upon a revolution

It is a somewhat crude simplification to suppose that science progresses from merely making ever more detailed observations. No matter how carefully you set up an experiment—and no matter how carefully you watch the outcomes—any observation has to be given an interpretation, and the interpretation you give can have an enormous influence upon the results. This is not to say however that science is an irrational activity or that rational debate is never possible. Scientists can change their minds and come to adopt different theoretical frameworks. It is always possible to consider different interpretations of the same data, and weigh up their various pros and cons. And while it may be possible to understand the falling cannonballs equally well in either Aristotelian or Galilean terms, it does not follow that one can hold either interpretation indefinitely. One of the many reasons why Aristotle's theories of motion are no longer serious contenders in the modern age is the fact that we can conduct experiments far beyond the scope of what was available in the sixteenth century. We don't need to drop cannonballs from the top of a tower, because we can go into outer space and drop all variety of heavy bodies to our heart's content. On the surface of the moon, where the effects of air resistance are negligible, different bodies of different mass will indeed fall at the same rate of motion, and we can see hammers and feathers falling—very slowly—in perfect accord. As we saw in the previous chapter, there will technically always be ways in which to preserve a theory in the face of recalcitrant evidence, but eventually, even the most committed of Aristotelians will have to concede defeat.

But what all this does show is that the outcome of a scientific experiment can often involve a complex series of negotiations. It is an attempt to find the theory that best explains the experimental data, and which also fits into everything else that we believe. In many cases, it is as much a matter of showing the plausibility of a particular interpretation as it is of demonstrating a new and unexpected result. Thus when Vincenzo Viviani wrote his story about Galileo and the Leaning Tower of Pisa, he was not simply misremembering an event that never happened or trying to spice up his biography with exciting anecdotes. It was rather a sustained attempt to generate support for the new scientific world view, a deliberate strategy for making the Galilean interpretation more plausible. And it worked. That is why the story continues to have resonance in the modern day, despite all of its glaring historical inaccuracies. Sometimes we make progress in science by conducting careful experimentation—but sometimes it can be just as important simply to tell everyone a damn good story.

This emphasis upon telling a good story also helps us to understand the larger context in which this whole incident took place. In 1543, a Polish mathematician and astronomer called Nicolaus Copernicus published a book entitled *De Revolutionibus Orbium Caelestium—On the Revolutions of the Celestial Spheres*—arguing that the Earth revolved around the Sun. This was in direct opposition to the prevailing orthodoxy which, following Aristotle, maintained that the Earth was stationary in the center of the universe, and that the Sun and everything else revolved around it. Over the years, this astronomical model had been articulated and developed into a highly complex system with enormous predictive power. It was known as the *Ptolemaic* system, after the Greek astronomer Ptolemy who did much to establish this framework in the second century. Make no mistake, it was an exceptionally good system, and after 1,400 years of refinement it provided an extraordinarily effective tool for navigation, calendar reform, and all of the other important tasks with which astronomy is concerned. But it was not perfect, and over the years there had been many suggestions for how it might be improved. Copernicus was far from being the first person to propose that the Sun should be transposed to the center to the universe and the Earth put in motion, but he was the first to formulate the idea in sufficient mathematical detail to command serious attention. For the first time in a millennium, there were two credible interpretations for the same astronomical data, and it was into this rhetorical battle that Galileo found himself drawn.

That the conflict between Ptolemaic and Copernican astronomy was a matter of interpretation rather than experimental investigation can be seen from the fact that they both predicted *exactly the same* observations. Both systems had been explicitly designed to accommodate the existing astronomical data. There was no scientific experiment—not even in principle—that could tell them apart. The issue ultimately came down to one of mathematical elegance. In the Ptolemaic system, every heavenly body is placed on a series of concentric orbits around the Earth, rings within rings. Those with the quickest orbit were placed closest to the Earth, beginning with the moon which takes roughly twenty-eight days to complete its circle, followed by Mercury, Venus, and then the Sun, which takes roughly 365 days to do the same. After that came the outer planets that were known at the time, Mars, Jupiter, and Saturn, the last of which appeared to complete its course only once every twenty-nine years. There were however some difficulties in this otherwise harmonious picture. The actual length of time taken by any of these planets to complete their orbit could vary considerably. Moreover, their motion often appeared to be far from uniform. Planets could speed up during certain parts of their orbit, appearing sometimes to be racing across the night sky, while at other times ambling sedately along their course. Their apparent distance from the Earth could also change dramatically over time, with some planets looming larger at certain points of their journey, all of which was somewhat difficult to explain given what was supposed to be a perfectly circular orbit. Worst of all, some planets could even appear to move backwards for brief periods during their orbit—the phenomenon known as *retrogression*—which was very disturbing indeed.

The Ptolemaic astronomers had therefore introduced a number of ingenious technical devices into their system in order to accommodate these apparent discrepancies. Orbits could be displaced, so that a wayward satellite would move in a perfect circle around a spot slightly off-kilter from the rest of the cosmos. This is called an *eccentric* orbit and was designed to ensure that at certain points during its course across the heavens, the planet in question will indeed pass closer to the Earth. Alternatively, the orbit of the planet remains focused around the Earth, but its rate of rotation is made uniform about a spot other than the exact center of the universe. This is known as an *equant* and is a mathematical device for explaining why, from our perspective on the Earth, certain planets appear to move at different speeds during their orbit. The most important device however was the idea of an *epicycle*. In the simplest cases, this is an orbit within an orbit—the planet in question moves in a smooth circular path about a point which is itself moving in a smooth circular path about the Earth. The overall effect will be a sort of corkscrewing motion, which amongst other things will result in the apparent retrogressive motion of the planet in question as it spins about its larger path through the heavens.

By the time of Copernicus, there were any number of different combinations of these different devices proposed for accommodating our astronomical observations. It is therefore slightly misleading to speak of *the* Ptolemaic system. Nevertheless, the various systems of the day all incorporated both eccentrics and equants, and about eighty different epicycles in some combination or other in order to make the calculations come out right. That is without doubt a great deal of mathematical complexity and geometrical fudging for an astronomical model that only contains seven moving parts, and might indeed raise the prospect that a different approach was required. Copernicus' insight was that all data requires an interpretation, and that sometimes the best interpretation can involve looking at things from the opposite perspective. By making the Sun the center of the universe, and putting the Earth in motion, it was possible to simplify the existing Ptolemaic system, and to eliminate many of the eccentrics and epicycles that made it look so unattractive.[5]

But there remained a problem. Simplicity can itself sometimes be a matter of interpretation. Certainly some of the mathematical techniques employed by Copernicus were ingenious and intellectually

FIGURE 2.1 *For the medieval astronomer, geocentricism offered both a satisfactory model of man's astronomical place in the universe, and his moral place in the universe. For just as crude matter is drawn towards the center of the Earth, so too is man dragged down by his corporeal sins; and just as fire is drawn upwards towards the outer planets, so too is the rarefied soul drawn upwards to the heavens beyond.*

pleasing in an abstract sort of sense. One could not help but admire the workmanship. But the resulting system still incorporated a good thirty-four epicycles of its own—an improvement over Ptolemy to be sure, yet still far too many fudges and fixes to convince anyone that it was any more likely to be true than its rivals. We are still only trying to accommodate seven moving bodies after all. Moreover, many calculations were much more difficult to perform within the Copernican system. That is why celestial navigation is *still* taught within a Ptolemaic framework, even though we now have absolutely no doubts at all as to the relative positions of the Sun and the Earth. But most importantly, if the Earth was indeed moving as Copernicus proposed, why did all the evidence of our senses suggest otherwise? We have all experienced the effects of traveling upon a rapidly moving surface. Things shake, coffee spills. And that is only at moderate speeds. If the Earth really was hurtling around the Sun at about 1,000 mph, then it seems that we should be experiencing these kinds of effects *all the time*.

This at least is the conclusion Aristotle would have drawn. On his account, recall, everything has a natural inclination to return to its proper place in the cosmos. All heavy objects move towards the center of the universe, the Earth included. It follows then that if the Earth was revolving around the Sun, it must be experiencing a constant force deflecting it from its otherwise preferred direction of travel, and it is these unnatural perturbations that will cause the constant, worldwide spillage of hot cups of coffee. In simple terms, the Aristotelian physics of the day could not comprehend the notion of something moving without the continuous action of a force—in modern terminology, it had no concept of *inertial motion*—and thus could not see how the Earth could orbit the Sun without literally being shaken to bits. In order to endorse the mathematical elegance of the Copernican system therefore, one also needed a completely revised account of the mechanics of motion. In order to make a small adjustment in astronomy, it turned out that one also needed to entirely rewrite the physics textbook. And that turns out to be an even better story.

Making the Earth move

The Copernican revolution was all about a change in perspective. By looking at things from the point of view of an Earth that revolved around the Sun—rather than an Earth lying stationary in the center of the universe—it was possible to see the movement of the planets in a whole new light. Such a change in perspective also allows us to better understand the curious story with which we began. Galileo never dropped cannonballs from the top of the Leaning Tower of Pisa, and part of the reason we can be so sure about this, is that we know that any such experiment would have been a complete and utter failure. Galileo wanted to undermine the Aristotelian theories of motion that made the prospect of a moving Earth so difficult to believe. But simply releasing these cannonballs from a great height would have done little to convince the scientific community to abandon its deeply held convictions. As we have seen, any such experiment could have been interpreted in all sorts of different ways. What Galileo needed was a sustained attack upon the fundamental principles of Aristotelian physics. It was only once such a project had been completed that such an experiment could be interpreted in the right sort of way. Our story about the episode at the Campanile therefore could never have been the *cause* of the scientific revolution. It was in fact one of its *consequences*.

Galileo's sustained attacked on Aristotelianism finally came in 1632, when he published his *Dialogue on the Two Chief World Systems*. The book takes the form of a discussion between three friends at a country estate, with Salviati advocating the merits of Copernicanism, Simplicio defending the traditional views of Aristotle and Ptolemy, and Sagredo—their wealthy host—acting as a kind of neutral observer. As we might expect, the *Dialogue* is as much an ingenious piece of literature as it is a penetrating scientific exposition. It is a very entertaining read, but there is no disguising its underlying message. Salviati is clearly intended to be Galileo's mouthpiece, and ends up running intellectual rings around his companions. Simplicio, whose name carries the same connotations of the country bumpkin in Italian as it does in English, frequently interrupts proceedings to make elementary arithmetical mistakes and generally display his ignorance. It is therefore no surprise that by the end of the *Dialogue*, Sagredo—standing in as a proxy for the reader—becomes convinced of the unequivocal superiority of Copernican astronomy.

The conversation ranges over a number of important topics, but the main focus is of course the movement of the Earth, and its apparent inconsistency with our everyday experience. Since Simplicio is apparently unable to even properly articulate this Aristotelian line of thought, Salviati states the argument as follows:

> As the strongest reason of all is adduced that of heavy bodies, which, falling down from on high, go by a straight and vertical line to the surface of the Earth. This is considered an irrefutable argument for the Earth being motionless. For if it made the diurnal rotation, a tower from whose top a rock was let fall, being carried by the whirling of the Earth, would travel many hundreds of yards to the east in the time the rock would consume in its fall, and the rock ought to strike the Earth that distance away from the base of the tower.[6]

The same underlying point is further expanded upon with other colorful examples. It is noted that if the ground beneath our feet were in constant movement as Galileo suggests, then one should be able to fire a cannonball considerably further in one direction than another—since it would be moving in the opposite direction to the ground—which is clearly not the case. It is also noted more simply that if the Earth and everything upon it were in a constant state of motion, the birds would be swept away and we should constantly experience a powerful wind in our face blowing from the east.

In very simple terms, the solution to these difficulties is to realize that if the Earth is moving, then everything else is going to be moving along with it. Galileo—sorry, Salviati—illustrates this by considering the case of a ship at sea. Suppose that the ship is moving at a constant speed towards the east, and that someone climbs to the top of the mast and drops a cannonball onto the deck. Since the ship is moving at the moment the cannonball is released, it will impart to it some of its forward momentum, in much the same way that a rapidly moving arm imparts forward momentum to a thrown projectile. It follows then that, from the perspective of a stationary observer standing on the shore, both the ship *and* the cannonball will appear to move slightly to the east during the time taken to complete the experiment, which is why the cannonball will still land directly below at the foot of the mast. From the perspective of the sailors however, who are also moving at a constant speed along with the ship, all they will see is the cannonball falling in a straight-line. The motion of an object therefore is not an objective matter, with everything heading towards a privileged spot at the center of the universe as Aristotle maintained. Rather, motion is relative to one's frame of reference. It is the same phenomenon that explains why, when watching Jean-Claude Van Damme's classic movie *Universal Soldier* on an aeroplane rather than working on your book and you manage to spill your coffee in excitement, it falls directly into your lap rather than hurtling towards the back of the cabin at 600 mph. The forward motion of the aeroplane is imparted equally to both the boiling hot coffee in your cup, and the pristine white chinos you are wearing on your seat, so that when one splashes towards the other, they continue to move at the same relative speed. From the perspective of a stationary observer in air traffic control of course, the coffee is moving across the horizon at an incredible rate of knots. It's just that everything else around it—including your trousers—is also moving in the same direction and at the same speed.

Shortly after the publication of his *Dialogues Concerning the Two Chief World Systems*, Galileo was summoned to Rome to defend himself against charges of heresy. It was alleged that his recent book contained a clear endorsement of the Copernican cosmology and was therefore contrary to the teachings of scripture. Amazingly enough, Galileo's cunning literary device of presenting the issue as a merely hypothetical debate in which the Aristotelian is repeatedly mocked and humiliated managed to fool nobody whatsoever. In fact, it seemed to have made certain members of the academic community—those who found their own arguments advanced by the ill-fated Simplicio—rather cross indeed. Words were spoken. Veiled threats were made. And so in the winter of 1633, an elderly Galileo found himself making an unwilling pilgrimage to the Vatican.

The whole incident caught Galileo very much by surprise. He was of course under no illusions that he had managed to make any number of enemies amongst his academic colleagues. But Galileo was a devoted Catholic, and had taken extraordinary pains to show that his scientific work was perfectly compatible with the teachings of the Church. Indeed, it was difficult to see what all the fuss was about. There are after all only a handful of passages in the Bible that seem to have even the remotest bearing on questions of astronomy, and even then they tend to be pretty inconclusive. Perhaps the most popular example is Joshua 10:12-13, where God is said to slow down the passage of the Sun in order to give the Israelites more time in which to defeat their foes:

> Then spake Joshua to the Lord in the day when the Lord delivered up the Amorites before the Children of Israel, and he said in the sight of Israel, "Sun, stand thou still upon Gibeon; and thou, Moon, in the valley of Ajalon." And the Sun stood still, and the Moon stayed, until the people had avenged themselves upon their enemies. Is not this written in the Book of Jasher? So the Sun stood still in the midst of Heaven, and hasted not to go down about a whole day.

I have no idea who the Amorites were, or why the Israelites were so pissed at them. The point presumably though is that God could hardly make the Sun stand still unless it was already in motion. Thus it follows that the Sun must revolve around the Earth, and not the other way around.[7]

But even leaving to one side some of the deeper theological issues concerning scriptural exegesis, it is clear that this is a pretty weak argument. The fact that the Sun appears to move across the sky is equally well explained by both Aristotelian and Copernican accounts. That's the whole point, since they are explicitly designed *to explain the same data*. It doesn't matter then whether or not the Sun revolves around the Earth, or the Earth revolves around the Sun, since either way the Israelites are still going to observe exactly the same phenomenon. In order for the Sun to stand still over the city of Gibeon, God is going to have to step in and temporarily suspend normal operations. That much at least seems clear. But there is nothing in the scriptures to indicate whether he does this by checking the motion of the Sun around a stationary Earth, or by halting the orbit of the Earth around an otherwise stationary Sun. To put it simply—as Galileo did himself on many occasions—there is no reason to suppose that Copernicanism clashes with scripture unless you are *already* committed to reading that scripture along explicitly Aristotelian lines. As with so much in this debate, it all comes down to a matter of interpretation.

Publish and be damned

The final episode of Galileo's life turns out therefore to be almost as mysterious as that with which we began. The facts about the Leaning Tower of Pisa simply do not add up. Nor do the facts concerning his trial and imprisonment by the Church. There was no particular scriptural interpretation that could clash with his scientific views, nothing in the Bible one way or another that could reasonably have any bearing on questions of astronomy.

But more importantly, it is also far from clear *why* such an interpretation was even an issue in the first place. The idea that scripture could have anything to say about scientific matters was in fact widely rejected amongst the ecclesiastical authorities of the day. Indeed, the dominant view was almost precisely the reverse—human reason and scientific investigation should in fact be used to help us interpret and better understand scripture, rather than the other way around. This was a tradition that stretched back to some of the most venerated Fathers of the Church. In a book arguing against the literal interpretation of the Book of Genesis, written in 415, St. Augustine stated that:

> In matters that are obscure and far beyond our vision, even in such as we may find treated in Holy Scripture, different interpretations are sometimes possible without prejudice to the faith we have received. In such a case we should not rush in headlong and so firmly take a stand on one side that, if further progress in the search for truth justly undermines this position, we too fall with it. That would be to battle not for the teaching of Holy Scripture but for our own, wishing its teaching to conform to ours, whereas we ought to wish ours to conform to that of Holy Scripture.[8]

It is a matter of humility. The world is a highly complex affair, and even our best scientific theories find themselves revised and rejected in the face of new evidence. Similarly, we should therefore not be too quick—nor so arrogant—to suppose that after centuries of dispute and debate, we alone have discovered the ultimate meaning of scripture. For St. Augustine as well, it is also a matter of priorities. The purpose of the Church lies in the salvation of souls, and time spent contemplating the inner mysteries of nature and the relative positions of the Sun and the Earth was—frankly—time that could be better spent attending to the poor and ministering to the sick.

There was therefore no serious or sustained theological opposition to Galileo's scientific work. And indeed, when he published his first serious attack on Aristotle in 1610, the Vatican responded by organizing a conference in his honor. Galileo had used his newly perfected telescope to show that the heavens were not perfect and incorruptible as tradition maintained. The moon was marked with mountains and craters, and little spots could be seen to swirl like clouds across the surface of the Sun. But while the Church celebrated his findings, Galileo's achievements and growing reputation did little to endear him amongst his academic colleagues, and 1610 also saw the beginning of a bitter and lifelong priority dispute with the German astronomer Christoph Scheiner over the discovery of sunspots (although ironically, neither Galileo nor Scheiner were the first to make these observations). Galileo's colleague at Padua, Cesare Cremonini, initially refused even to look through the new-fangled telescope as a matter of principle. He maintained that astronomy was the business of learned philosophical speculation, and not the grubbing around with bits of wood and mirrors like some common tradesman. He was forced to retract his views after a humiliating public display at the university. Ludovico delle Colombe, a professor at Florence, was so incensed at the attack upon his work that he even organized a hate campaign against Galileo specifically devoted to defaming his reputation. It was in fact this group of *scientists* that first raised the issue of scriptural compatibility against Galileo in what can only be described as a deliberate and cynical attempt to co-opt the moral authority of the Church in a fight to preserve the scientific consensus against damning new evidence.[9]

The issue came to a head in 1616, when the Church was finally forced to step in. Galileo was advised by his friends and supporters in the Sacred Congregation of the Index—the literary branch of the Holy Inquisition, a sort of 1,001 Books You Should *Not* Read Before You Die—that he should no longer *endorse* the truth of Copernicanism, but that he could continue to discuss the theory as a matter of mathematical interest. Galileo happily agreed, and his opponents were temporarily silenced. But when in 1632 the bumbling Simplicio cheerfully reproduced the arguments of Cremonini and Colombe only in order to be quashed by the unbearable Salviati, old wounds were reopened. At which point events took a very strange turn indeed.

When Galileo arrived in Rome, he was charged not with teaching a theologically suspect account of the motions of the Earth, but of the much more serious offence of deliberately defying the Pope. A document had been produced stating that in 1616 Galileo had been instructed not only to refrain from endorsing Copernicanism, but also in fact to never again even *discuss* the theory on pain of imprisonment. Needless to say, this came as something of a surprise to Galileo, who had been working under the assumption that Copernicanism could still at least be presented as a mathematical hypothesis. Indeed, he still possessed the signed edict from the Sacred Congregation of the Index confirming this decision, which he duly presented to the court. To make matters even worse, his *Two Dialogues* had even been officially approved by the Holy Inquisition before going to press, the documentation for which Galileo was also able to produce. One can imagine the awkward coughs and whispered arguments. The trial was adjourned for the rest of the day.

As it turned out, the incriminating document that had caused all the trouble—the one stating that Copernicanism could not even be discussed—also demonstrated a number of legal irregularities. In particular, it had never been signed, neither by Galileo nor by any member of the Inquisition involved with the original dispute. In all likelihood, it was nothing more than a rough draft drawn up by a courtroom clerk and later superseded by the actual edict that Galileo had produced. Quite how this document had survived is a mystery. Quite how it had found its way into the highest circles of the Holy Inquisition is a matter of continuing speculation. Yet of all the various factions who might have held a grudge against Galileo, there was none more embittered—or for that matter, better placed to influence the operations of the Inquisition—than the Jesuits, who for centuries had enjoyed enormous prestige as the leading scientific authorities of the Church and stalwart defenders of the Aristotelian orthodoxy. Indeed, their most eminent astronomer was none other than Christoph Scheiner himself, who even made a guest appearance at Galileo's trial in order to gripe about his old priority dispute.

After extended legal wrangling, it was finally decided to offer Galileo the equivalent of a plea bargain. He would confess to a minor charge—an overzealous presentation of Copernicanism that might be misconstrued as an endorsement. A slapped wrist, some minor edits to *Two Dialogues*, and everyone saves face. Yet somehow things didn't go to plan. When the paperwork finally made its way to the Pope for official sentencing, a very different story presented itself. The transcripts for the trial bore absolutely no resemblance to what had actually happened. There was no mention of the plea bargain, or the conflicting edicts. Somehow or other these seemed to slip the mind of the cleric—who just happened to be another Jesuit—in charge of recording the court proceedings. A combination of careful editing and straightforward fabrication presented a defiant Galileo in utter contempt of ecclesiastical authority. The Pope was furious. Galileo was sentenced to a lifetime's imprisonment, later commuted to permanent house arrest, where he died nine years later in 1642.

Concluding unscientific postscript

In May 2015, Pope Francis issued his second encyclical *Laudito si'*—*Praise Be to You*—and was widely hailed in secular quarters for ushering in a new, progressive attitude within the Roman Catholic Church. For while the document clearly reaffirmed the Church's traditional opposition to issues such abortion and stem-cell research, it nevertheless gave a clear and unequivocal endorsement to the reality of manmade climate change and the pressing need to reduce fossil fuel emissions. Coming nearly 400 years after Galileo was supposedly tortured for climbing the Campanile and pointing out the obvious, the encyclical has been seen as the Church's willingness finally to accept the scientific consensus, rather than opposing it on the grounds of religious authority or orthodox scriptural exegesis.

The problem of course—as we have seen—is that the idea that Galileo was persecuted for heretical beliefs simply does not add up. His views were acceptable, celebrated by the Vatican, and there was a general agreement that science and scripture were completely compatible. Galileo's real enemies were not clerics and theologians, but *other scientists*—individuals humiliated by his break with the scientific consensus, and who enjoyed sufficient influence with the religious authorities to exact a terrible revenge. Through an exacting process of trial and error, Galileo discovered that a scientific consensus need not always concern itself with minor issues like empirical evidence and mathematical elegance when part of its interpretation involves the much more important element of commanding the moral high ground instead.

The reality then is that Pope Francis was not really breaking any new ground when he publicly endorsed the compatibility of science and religion, and in this respect the comparison with Galileo is a particularly unfortunate one. The Church has long been closely involved with the development of science, and has frequently found itself at the forefront of defending the scientific consensus. What the story of Galileo really shows us then is that this does not always have a happy ending.

3

Images of science

It is probably not too much of an exaggeration to say that Isaac Newton was one of the greatest natural scientists who ever lived. Born the son of a farmer in Grantham in 1642, and initially supporting himself through his undergraduate education working as a gentleman's valet, by the age of twenty-seven Newton had already been appointed to the Lucasian Chair of Mathematics at the University of Cambridge, and was well on his way to developing the theories of mechanics and gravitation for which he is still famous. His most important work, *Philosophiae Naturalis Principia Mathematica—Mathematical Principles of Natural Science*—was published in 1687, and in many ways marked the culmination of the scientific revolution that had begun with Copernicus and Galileo roughly 100 years earlier. What Newton showed was not only that the theory of motion developed by Galileo could be reconciled with the heliocentric model of the solar system proposed by Copernicus, but that they could in fact both be explained in terms of the same basic principles. That is to say, the underlying physical laws that determine the way in which (say) an apple falls from a tree are in fact *the very same* physical laws that also determine the way in which the Earth orbits the Sun. It was a staggering result, another nail in the intellectual coffin for those who still maintained that the heavens above were governed by their own unique laws. Newton's *Principia* was therefore without doubt a remarkable piece of intellectual synthesis. It was also a technical masterpiece of such novel mathematical complexity that practically no one else on the planet could understand it—but that of course only further enhanced its reputation.

Needless to say, such advances in our scientific understanding did not come without their fair share of difficulties. In order to properly express his new picture of the world, Newton had to first of all invent a whole new branch of mathematics capable of handling the sophistication of his ideas. These were the principles of calculus that we all now know and love, and Newton considered it one of the crowning achievements of his career. He was therefore understandably annoyed when the German philosopher Gottfried Leibniz suggested that maybe he had invented it first, and that Newton had actually plagiarized the whole thing. Angry letters were exchanged, and much public condemnation followed in the press. Eventually the Royal Society intervened, and in 1712 published their conclusions, finding unequivocally in favor of Newton. Lest there be any suspicion of partisanship or bias, it should be noted that the President of the Royal Society himself—who (cough) just happened to be Isaac Newton—oversaw the entire investigation from beginning to end, and personally wrote the final report.

A second difficulty was that while Newton had managed to formulate the laws of gravity with an unprecedented degree of mathematical rigor, he still had absolutely no idea *what exactly* it was that he was supposed to be formulating. Most physical interactions have a reassuringly concrete basis—one billiard ball bangs into another billiard ball, and off they go towards the far end of the table. But gravity is not like that. In fact, it doesn't seem to involve any kind of physical interaction at all. The planets of the solar system just seem to be held in place by some kind of mysterious and invisible force that operates over vast distances of empty space with absolutely nothing in between. Critics therefore complained with some justification that Newton had simply reintroduced those very same occult qualities that the scientific revolution was supposed to have overthrown. For the new generation of modern scientists, the universe was supposed to work like clockwork, but without a physical mechanism underlying the force of gravity, none of the cogs and springs meshed together. Newton replied that it was God himself who intervened in order to make sure that the whole of creation kept moving in the right sort of way; Leibniz quipped that it was a disappointing deity who was forever having to stop to wind-up his watch. But then again, Newton had a number of unconventional views about religion. He probably devoted more of his life searching for hidden messages in the Bible than he did studying physics, and while he refused to indulge in the usual speculations regarding the date of the coming apocalypse, he nevertheless proved mathematically that the earliest possible date for the end of the world was in 2060, which works just fine for me.

But by far the most serious difficulty that Newton faced, however, concerned the accumulation of sufficient data. If you are going to provide the basic mathematical principles governing every single physical interaction in the known universe, you are clearly going to want to work with the widest possible range of evidence available. One of the centerpieces of Newton's account was to show how the gravitational attraction of the Moon on the Earth could even explain the motion of the tides—a problem that had defeated Galileo, and which had hung over the subsequent scientific revolution as something of an embarrassment.[1] Newton therefore began dutifully to compare his calculations with the records and reports of sailors, fishermen, and pilots throughout England's vast trade network. Unfortunately for Newton, many of the professional seafarers involved in the project understood all too well the economic realities they faced if all of their years of experience in the subtle arts of navigation could be reduced to a handful of calculations reproduced in a book. What Newton received therefore was in fact a wealth of bizarre and mutually contradictory reports that systematically exaggerated the danger and unpredictability of many of the world's most benign shorelines—not to mention the heroic self-sacrifice and utter indispensability of those who continued to battle these capricious currents on a daily basis.

While some of these difficulties were eventually resolved through the expedient device of sending out an enthusiastic amateur from the Foreign Office with his own tape measure and pocket watch, Newton would still find himself going to extraordinary lengths in his quest for reliable data. In one of his notebooks from 1665, he records an experiment he performed in order to investigate the inner workings of the eye. It was part of Newton's mechanical world view that all physical phenomena could ultimately be explained in terms of the interaction of smaller moving parts—again, as in some giant piece of elaborate clockwork—and so he conjectured that vision itself must be somehow due to the collision of little particles of light impressing themselves upon the surface of the eye. The difficulty, however, was finding a way to test the hypothesis, since merely dissecting an eyeball and fiddling around with its parts could tell us very little about the subjective experience of these interactions. Newton's solution was extremely straightforward:

> I took a bodkin and put it betwixt my eye and the bone as near to the backside of my eye as I could: and pressing my eye with the end of it ... there appeared several white, dark, and coloured circles. These circles were plainest when I continued to rub my eye with the point of the bodkin, but if I held my eye and the bodkin still, though I continued to press my eye with it, yet the circles would grow faint and often disappear until I removed them by moving my eye or the bodkin.[2]

A bodkin is a kind of knitting needle by the way, and basically what Newton discovered is that when you ram one of them into your eye socket and wiggle it around, you see all sorts of different colors. As I said, probably one of the greatest natural scientists who ever lived.

Unfortunately, for all of his extraordinary efforts, Newton was still to end up the victim of insufficient data. The problem was that he restricted the range of his investigations to an extremely parochial set of concerns—namely, things that directly affect human beings on the planet Earth. This might have seemed like a reasonable decision to make at the time, but as Einstein showed at the beginning of the twentieth century, things can begin to look very different once we move to the cosmic scale. Some of the fundamental constants of Newtonian Mechanics, like the mass of an object and the passage of time, begin to behave very differently as we approach sufficiently high velocities. Of course, most of these factors are simply undetectable at the human scale. It requires a supersonic jet and an atomic clock of mindnumbing precision in order to record the most minuscule effects of time dilation. You can put a man on the moon without having to worry about the effects of relativity. And indeed, when the velocities in question are sufficiently small, it is in fact possible to derive Newton's original equations as a limiting case from the more complex framework provided by Einstein. But while Newtonian Mechanics may well work for many intents and purposes, that still does not stop it from being *technically false*—an unwarranted extrapolation made on the basis of insufficient amounts of data.

How big's your data?

The case of Newton and his theories of mechanics and gravitation represent one of our most enduring images of the scientific method—the isolated individual going to extraordinary lengths to produce the necessary data to test their conjectures. But it is an image that has apparently become increasingly outdated as science has become an increasingly professionalized activity, with advanced experimental techniques, the rapid dissemination of results and global research networks encompassing countless individuals and laboratories. Indeed, while Newton may well have been forced to take some rather extreme measures in his attempts to scrape together enough data, the contemporary scientist by contrast seems almost spoilt for choice. We live now in an age of apparently abundant data, with better instruments, in more locations, recording events of an unimaginable breadth and variation. To be sure, not all of this information is terribly useful—I believe the collective term for this is "the Internet"—but even something as seemingly inconsequential as your browser search history now holds out the prospect of fueling the next great scientific breakthrough. If only Newton had access to the quantities of data available today, and of course the software necessary for analyzing it, he would never have made the same mistakes.

This at least was the theory behind Google Flu Trends, a mighty piece of data crunching software that sought to track infection rates across the U.S. purely on the basis of what people searched for on the internet. In a paper published in the prestigious scientific journal *Nature* in 2009, Google reported that its results very closely matched the data recorded by the Centers for Disease Control and Prevention (CDC)—which worked on the more traditional model of simply recording the number of people who visited the doctor with a nasty case of the sniffles—but much quicker and much more efficiently. For whereas CDC data was always a few weeks behind any flu epidemic, since it obviously enough had to wait for people to actually get sick before it could record their numbers, Google Flu Trends could provide up-to-the-minute data on internet search results. To search for a particular inelegant neologism, it wasn't just *forecasting* the spread of infection based on previous data, it was *nowcasting* the infection in real time. And all of this without having to build complex computer models or thinking long and hard about the potential infection mechanisms. All you had to do was shovel the data into an algorithm and start cranking the handle. It was heralded as the Big Data revolution—a whole new era in scientific research of fast-paced statistical analysis without all of that tedious theory slowing it down. It was in short an entirely new image of science. Newton's problem had finally been solved.[3]

Unfortunately, as the foregoing tincture of sarcasm might have suggested, the honeymoon period did not last very long. The following year, in 2010, good old CDC data started to outperform Google Flu Trends. It was found that if you simply extrapolated in a straight line from the last two weeks of reported doctor's appointments, you got much better predictions about the future rate and spread of infection than anything that the multimillion dollar search provider could produce. In 2012, Google Flu Trends was overestimating the number of infections by about 50 percent. By 2013, it was overestimating by 100 percent—that's twice as many infections as was actually the case. By that point, you could make more accurate predictions about how the infection would develop by simply taking *last year's* data and assuming that everything was just going to happen exactly the same way as it did before (that is to say, by repetition rather than prediction). In 2014, Google Flu Trends was quietly discontinued. *Vive la Révolution!*

So what exactly went wrong? It is unfortunately a little difficult to provide a precise diagnosis of Google Flu Trends, since Google itself is understandably very protective about the inner workings of its fabled search algorithms. But one initial observation that we can make is that more data does not always mean better results. This is one of the many lessons that have been learnt from opinion polls. At the time of writing this, many political pundits in the U.S. are still reeling from the election of Donald Trump to the White House, a result widely and confidently predicted by any number of pollsters and data crunchers to be completely impossible. This is, however, far from being the biggest upset in U.S. political history. For an even better example, consider the case of the 1936 U.S. presidential election, between the democratic incumbent Franklin D. Roosevelt and his Republican opponent Alfred Landon. While opinion polls were of course already well established by the time, this competition witnessed a staggering increase in enthusiasm, with the *Literary Digest* polling an unprecedented 2.5 million people in an attempt to determine the outcome. The results were conclusive, with Landon predicted to win 51 percent of the vote. In the end of course, Roosevelt romped home in one of the largest landslides in U.S. history, winning over 98 percent of the electoral college votes—the largest number since James Monroe ran unopposed in 1820—and over 61 percent of the popular vote, a feat that surpasses even Ronald Reagan's crushing victory in 1984. The *Literary Digest* wasn't just wrong. It was spectacularly wrong. Categorically and catastrophically wrong. And just to make matters even worse, a much smaller poll of around 3,000 people conducted by *Gallup* got the result almost spot-on. As the old saying goes, size isn't everything.

The problem of course was that while the *Literary Digest* had a much larger sample of people, it wasn't a very *good* sample. It was not representative of the population as a whole. In particular, the names for the poll had been compiled from a number of easily available lists—such as telephone directories and automobile registrations—which back in 1936 only covered a very selective cross-section of voters, specifically the more prosperous members of society who were already more likely to vote Republican. And this is hardly an isolated incident. More recently in the United Kingdom for example, opinion polls have tended to systematically overestimate support for the Labour Party, as demonstrated by the surprise Conservative victories in 1992, and again in 2015. This is widely attributed to what U.K. pollsters refer to as the 'Shy Tory' syndrome—the idea (as presumably suggested by Labour voters) that since those who vote for the Conservative Party are immoral corporate fat-cats, far more interested in giving tax breaks to their chums in the banking sector than they are in serving the public interest, they tend to be somewhat reticent about *admitting* their voting habits in public.

This should presumably be contrasted with the 'Tiresome Liberal' syndrome—the idea (as presumably suggested by Conservative voters) that since those who vote for the Labour Party are self-important narcissists, far more interested in securing dinner party bragging rights for supporting a supposedly progressive party than they are in offering coherent economic policies that might actually help people, they tend to *never shut up* about their voting preferences, whether you poll them or not.

Thus just because Google Flu Trends was able to analyze vast quantities of data, it doesn't follow that its conclusions were going to be any more reliable than those the CDC could offer. But there is more. A second possibility that has been suggested is that Google Flu Trends failed to account for the influence of advertising and the media. If the newspapers are full of sensationalist stories of killer flu and sweeping epidemics, or if the local pharmacy is running a particularly aggressive advertising campaign for flu shots, then people are far more likely to search for some of the terms flagged by Google Flu Trends. And, of course, there is no guarantee that the extent of media coverage will be in any way proportional to the actual severity of the flu season. There is no quicker way to fill a slow news day than to track down a sick pensioner and rehash last year's dire warnings about the new strain of drug-resistant foreign super-viruses, spreading across the border, coming over here and taking our jobs ... Eventually even the healthy among us start to wonder about that ache in our neck and that soreness in the throat, and the next thing you know you've self-diagnosed yourself with some unpronounceable tropical disease that up until now only ever affected parrots.

A third, and more amusing, possibility is that Google's own search engine might have managed to distort the data that was being tracked. Again, some of this involves a degree of speculation as to how exactly Google's algorithms operate, but it seems pretty clear that the results of one search can influence the results of another. Google often recommends "related searches" for the terms you have put in—which may or may not be determined by advertising revenue—and updates its recommended searches on the basis of other people's search history. It is therefore easy to imagine a situation where someone made a nonflu-related search, was randomly recommended something related to coughs and sneezes and on a whim, clicked the link. Perhaps someone was researching the movie *Double Impact*, the one where Jean-Claude Van Damme plays long-lost identical twins brought together in a quest for vengeance, and is offered a link to a particularly powerful dual-purpose decongestant with the same name, and decides to find out more. The immediate consequences of this are of course vanishingly small, but nevertheless just enough to ever so slightly increase the chances of Google recommending coughs and sneezes for the next nonflu-related search. And as the chances increase, more people click the link, and the process begins to snowball. Suddenly lots of people are searching for flu symptoms, and for no other reason than that Google *itself* is encouraging them to search for flu symptoms. Ironically enough, the way in which a search term can replicate and spread throughout the internet is itself probably an excellent model for the way in which real-world viruses can spread, with the physical proximity of infected individuals in a society replaced by some more abstract measure of mathematical proximity of terms within an algorithm. But as fascinating as this might be, it sadly has nothing to tell us about when actual people get ill.

 The real question then is not so much why Google Flu Trends began to go awry, but rather why it was ever successful in the first place. We have already encountered the problem of *sample bias*. This happens when the data under consideration is not in fact very representative of the phenomena you want to study. Most opinion polls face the problem of sample bias in some form or other, whether in the extreme case of the *Literary Digest* only polling those in 1936 wealthy enough to own a telephone or automobile, or the more modern problem of certain individuals being more willing to announce their political preferences than others. And while the Big Data revolution does not guarantee the elimination of sample bias—how much data is enough?—it is nevertheless the case that this is always going to be a more significant problem for small data sets. But closely related to all of this is the problem of *confirmation bias*. This happens when we unwittingly give greater prominence to the data in favor of a hypothesis than to the data against it. A lot of superstitions are based on confirmation bias, since we tend to remember the one lucky coincidence far more vividly than the countless other instances when things didn't go so well. Or consider homeopathic medicine, which will tend to 'cure' your cold in about three days because all colds tend to clear up in three days regardless. And clearly enough, when you are dealing with the quantities of data involved in something like Google Flu Trends, the more likely it is that you will be able to spot *some* correlation that initially seems to confirm your hypothesis.

It's what you do with it that counts

Google Flu Trends was always going to be an answer to the wrong question. Newton's problem was never simply that he didn't have enough information, and the solution was never going to be finding ways to crunch ever more quantities of data. Indeed, imagine for a moment that Newton had been able to access a far greater number of reliable informants about the motion of the tides, or the swinging of pendulums, or any other of the raw facts and figures upon which he had constructed his theory. Suppose that Newton had had access to the sort of quantities of data that we have today, with real time information of sea levels at every point on the globe and the position of the moon, and the formidable Big Data algorithms necessary for shifting through those countless data points. The fact of the matter is that he would still have been wrong, and that is because the circumstances in which Newtonian Mechanics can be shown experimentally to break down—such as when velocities approach the speed of light—were circumstances that he did not even know he had to consider. The problem was never that Newton didn't have *enough* data. The problem was that Newton didn't have the *right sort of* data.

The problem is actually a perfectly general one, because if you think about it—and this really is the sort of thing that philosophers spend a lot of time thinking about—it's hard to see how we could ever have enough data to solve the sort of problem with which we began. Let us take a toy example, much beloved by the professional philosopher. Suppose that we are attempting to investigate the plumage on different species of bird. We observe a number of black ravens, and tentatively propose the conjecture that all ravens are black. But how many black ravens do we need to observe before we can be confident in our simple ornithological theory? At a first pass, we can see that it will depend upon how many ravens exist in the world. To take one extreme example, if there have only ever been ten individual ravens that have existed in the whole of creation, and we have seen eight of them and they have all been black, then our confidence can be pretty high. But to take another extreme example, if there are billions upon billions of ravens in the world, and we have only seen eight of them, then a considerably more cautious attitude is advisable. But the problem is not simply that the number of ravens is likely to be extremely large. That at least would allow us to make a reasonable (albeit pessimistic) assessment of the status of our theory. The problem is that we have *absolutely no idea* how many ravens there are that we have yet to observe—just as Newton had absolutely no idea that the range of data that he was considering was in fact limited to a terribly narrow range of parochial concerns. Indeed, if we were actually in a position to know roughly how many ravens there were in the world and how many more we needed to observe, we would probably already know enough about the raven population in general that we wouldn't have found it necessary to start speculating about their color and proposing such scientific theories in the first place.[4]

It seems then that we are never in a position to judge the reliability of our data, or at least, not in those situations of genuine interest to the practicing scientist. Nevertheless though, it does not seem unreasonable to suppose that the more instances we have that confirm our theory, the more likely it must be that our theory is true. The more ravens we observe, the more likely it is that all ravens are black—even if we cannot say for certain *how much* more likely it is that all ravens are black. But there may be a problem with even this modest assessment. Suppose that we have indeed observed a great number of black ravens (and only black ravens), and have therefore conjectured that every other raven must also be black. Part of what this conjecture involves is a prediction about the future. It is an extrapolation from what we have seen in the past to a claim about what we will see in the future, and any such extrapolation must make a number of assumptions about the behavior of the world. Specifically, any attempt we make to predict the future by extrapolating from the past must presuppose that the world is largely *uniform* in its behavior, and that things will more or less go on in the same way as before. We can see this when we consider that if the world was not largely uniform—and if things did not continue in more or less the same way as they did before—then it wouldn't matter how many positive instances we might have observed. If everything could change at any moment, then no number of black ravens could tell us a damn thing about the color of the next raven that we observe. In the absence of this crucial presupposition, we would have no way of determining whether we had been gathering evidence for the (uniform) assumption that all ravens are black, or whether we had been gathering evidence for the (nonuniform) assumption that all ravens have been black up to today but will be white tomorrow, or that all ravens have been black up to today but will be white next week, or that all ravens have been black up to today but will be white in a month's time, and so on and so forth. And the kicker is that this crucial presupposition—the claim that the world is largely uniform—is in fact just as problematic as our original theory.

This at least was the claim by David Hume, one of the leading figures of the Scottish Enlightenment, and arguably one of the greatest philosophers of all time. Hume's argument is deceptively simple, although its consequences are devastating. There is certainly no *guarantee* that the world is uniform. There is nothing logically incoherent in supposing that the world is radically unstable, or that every observed raven is going to suddenly change its color next week. It is not a matter of definition in the way that all triangles have three sides or that all bachelors are unmarried. It is rather something that we have to go out and discover for ourselves. But if the uniformity of the world is just another empirical fact, how could we go about establishing it? Like any other scientific conjecture, it is all a matter of gathering data. We would have to begin with a limited set of observations, and extrapolate to a more general conclusion. We observe that the world has demonstrated a reassuring degree of uniformity in the past, and we infer that it will continue to display the same degree of uniformity in the future. But now we are just running in circles. As Hume put it:

We have said that all arguments concerning existence are founded on the relation of cause and effect; that our knowledge of that relation is derived entirely from experience; and that all our experimental conclusions proceed upon the supposition that the future will be comfortable to the past. To endeavour, therefore, the proof of this last supposition by probable arguments, or arguments regarding existence, must be evidently going in a circle, and taking that for granted, which is the very point in question.[5]

The practice of predicting the future by extrapolating from the past is known more formally as induction, and the difficulty raised by Hume that we can never have enough data is similarly known in philosophical circles as the problem of induction.

As the reader might imagine, there remains something of a cottage industry still devoted to trying to resolve this issue. One extreme solution to the problem of induction of course is to simply abandon the principle altogether. If we cannot justify those inferences that seek to extrapolate from the past in order to make predictions about the future, then we should stop making such inferences and try to find some way of understanding the scientific method in their absence. This would take us back to the view of Karl Popper, who as we have already seen, attempted to reduce the entirety of the scientific method to nothing more than the process of falsification. Indeed, in addition to his desire to provide a clear demarcation between the open-mindedness of scientific enquiry and the dogmatic intransigence of pseudo-scientific nonsense, Popper was in fact also greatly troubled by the problem of induction, and saw it as a virtue of his theory of falsificationism that it offered a way to finesse this difficulty altogether. The idea presumably was that while we may never be able to know whether a well-confirmed theory today will continue to work in the future, we can presumably know that once a theory has been falsified then that is the end of the story.

We have of course already seen some of the difficulties in Popper's account, not least the fact that it proves to be remarkably difficult to ever find a clear and definitive instance of falsification. But there is yet another shortcoming in Popper's account relating to this exclusive reliance upon nothing more than the process of conjecture and refutation. Suppose that we have two competing scientific theories describing a single domain of enquiry, such as whether or not a particular bridge is going to be able to take a certain weight, or if a particular brand of medicine is safe to use, or something similar. And let us also suppose for the sake of argument that we have managed to definitively falsify one of these theories while the other remains an open question. And now imagine that we are facing a situation where we need to choose between these two different theories to help determine our future course of action—such as whether or not to drive over the bridge, take a medicine, or whatever the case happens to be. The sensible course of action is to rely upon the theory that has not yet been falsified. We would presumably reason to ourselves that we don't know whether or not that theory is true, but we do know that the other theory is false, and so we are simply choosing between possible success and certain failure. That would certainly be Popper's advice. But what exactly underpins such an assessment? Why should we suppose that the falsified theory will continue to be unreliable? What guarantee do we have that things will continue as they have done before, and that the past failure of one theory is a reliable guide to the future? It turns out then that the entire process of falsificationism must also presuppose that the world is largely uniform, since otherwise the fact of falsifying a scientific theory would be a meaningless achievement. Rather than finessing the need to make an inductive inference, Popper's theory of falsificationism in fact presupposes it.

A second popular response to the problem concedes that we can never know for sure if extrapolating from the past is going to work in the future, but nevertheless maintains that if there are *any* reliable methods for navigating the world around us, then induction must be among them. The idea is that we might just have to accept that the world is fundamentally random and unpredictable, in which case we are—scientifically speaking—completely stuffed whether we continue to reason inductively or not. But suppose on the other hand that there does exist some method for making reliable predictions about the future, even something that we would not usually consider to be part of the scientific method, such as reading tea leaves or consulting a crystal ball. No matter how strange and exotic the method, if there is some reliable method for predicting the future, then we can also trust the principle of induction. This is because of what it means to say that reading tea leaves or consulting a crystal ball is a reliable method of inference—namely that they have worked well in the past and that we expect them to work well in the future. In short, to say that some method of inference is reliable *is itself* to make an inductive inference; you can't have one without the other.

What we have then is not so much a justification of induction, but a cunning way of hedging your bets. Whatever way the world turns out to be, we might as well reason inductively. If the world turns out to be utterly random and unpredictable, we haven't lost anything since any method of inference was also going to fail. But if the world does turn out to be predictable, even in spooky and unexpected ways, then induction will find a way to piggyback on that success. It is definitely a philosopher's solution—intellectually pleasing and sophisticated, but somehow doesn't quite manage to scratch the itch. The problem of course is that it is simply too abstract. What we really wanted to know was whether or not any particular scientific theory was likely to be true on the basis of its past success. We wanted to know if observing black ravens was a reliable method for determining whether or not all ravens were black; Google wanted to know if recording internet search histories was a reliable method for plotting the spread of infection; and Newton wanted to know if measuring the motion of terrestrial phenomena was a reliable method for understanding the entire cosmos. All that the above reasoning tells us, however, is that *if* observing black ravens is a reliable method of speculative ornithology, *then* the inductive inferences we make on the basis of those black ravens will indeed be justified. But that is not so much answering the question as simply restating it.

Hume's own solution to the problem was a little more cynical. He notes that since we just seem to be hardwired to reason inductively anyway regardless of the philosophical difficulties involved, we should just relax and get on with it. For Hume then, science is quite literally an irrational activity—but maintains that the quest for rationality might well be overrated anyway. That is not to say that science is impossible or that we can never make steady progress in our investigations. But it does suggest that attempting to uncover a single unequivocal scientific method and using it as the standard against which other rational activity is measured is probably wrongheaded.

There is no brute force solution to Newton's problem. Once we start thinking of the scientific method as nothing more than a process of amassing ever more quantities of data—as simply a matter of size—then we have set ourselves a challenge that we cannot hope to win. And that sounds like a pretty good argument for looking for a better image of science.

Bad first impressions

Ultimately then, the problem of induction remains unsolved. And while this may be of some relief to the professional philosopher—after all, we all need something to do with our time—the result is considerably less satisfactory to the self-reflective scientist, who would presumably like some reassurance that the common-sense practice of gathering as much data as possible for their theories was not in fact an utterly futile activity.

It is worthwhile therefore spending a little time trying to diagnose how we managed to get ourselves into this predicament in the first place. The point is that there is a difference between gathering data that confirms our scientific theory—more black ravens—and gathering data that helps to establish *why* the scientific theory in question is true. And the root of the problem is that while we can easily observe more and more instances of our scientific theory, we can never observe the underlying link or necessary connection that guarantees that our theory will continue to hold in the future. Or to put it more simply, while we can observe *that* a particular raven is black, and make our extrapolations into the future accordingly, we cannot observe the fundamental link between being a raven and being black that guarantees that these extrapolations are going to hold.

Another way to think about the problem is that while we can easily enough observe the *correlation* between different events, it is considerably more difficult to observe the *causation* between different events. For two events to be correlated just means for them regularly to happen at more or less the same time. But for two events to be causally connected implies a more robust form of stability. It can sometimes be difficult to distinguish between the two. We know for example that whenever the little hammer in the church tower strikes the bell, it makes a noise. There is a straightforward causal mechanism to explain this, involving the vibration of the bell and the molecules of the air. More importantly, if the hammer had not struck the bell, there wouldn't have been a noise. By contrast, we also know that whenever the clock in the church tower strikes twelve, the clock in the town hall does the same a few seconds later. But that of course does not mean that the clock in the church tower striking twelve *causes* the clock in the town hall to strike twelve. While both clocks might well have been calibrated to the same standard (to within a few seconds' accuracy), there is nothing about the striking of the first clock that determines the striking of the second clock. In particular, if the clock in the church tower is removed for repairs, the clock in the town hall will carry on as before as if nothing had happened.

Similarly, in 2009 there was a high degree of correlation between internet search results for coughs and sniffles and the actual spread of infection across the U.S. But ultimately there turned out to be no causal connection between the two—there is no underlying, fundamental link between what you type into your browser and whether or not you get ill—which is why Google Flu Trends became more and more unreliable as time went on. What we really want then if we want to solve the problem of induction is to make sure that all the data we gather is in fact causally relevant. And this is precisely the problem. As Hume put it:

> When we look about us towards external objects, and consider the operation of causes, we are never able, in a single instance, to discover any power or necessary connection; any quality, which binds the effect to the cause, and renders the one an infallible consequence of the other. We only find, that the one does actually, in fact, follow the other. The impulse of one billiard-ball is attended with motion in the second. This is the whole that appears to the *outward* senses. The mind feels no sentiment or *inward* impression from this succession of objects: Consequently, there is not, in any single, particular instance of cause and effect, any thing which can suggest the idea of power or necessary connection.[6]

We see one billiard ball hit another, and the second ball move away. We can look closer and closer, right up to the moment of impact, but no matter how closely we look, we never actually see the first billiard ball *causing* the second billiard ball to move. At one moment it is stationary, and at the next it is rocketing off towards the top cushion. But no time-lapse photography will record the moment when the causation happens. We can examine the billiard balls to any degree of accuracy, measure their momentum and the transfer of kinetic energy, the quantum excitation of their individual atoms, and specify to the tiniest fraction of a second the point at which one billiard ball stopped moving and the other began—but we will never be able to observe the moment at which one billiard ball *caused* the other to move. Indeed, as Hume goes on to note, if we *could* observe the causal connection between events, we would know with certainty how the future would unfold, and there would be no need for science in the first place.

Underlying Hume's problem here is a particular conception of how we interact with the world. For Hume, observation is a purely passive activity—we open our eyes, and information simply floods inside. These simple ideas with which we are immediately acquainted Hume calls *impressions*, and it is out of these that we construct more complex ideas and abstract images when engaged in more theoretical reflection. It is a picture which makes the problem of induction especially intractable, since, as we have seen, the causal connection between events is not something that simply presents itself for our inspection. It is a picture exemplified with particular clarity by Newton, whose own experiments with a knitting needle were explicitly premised on the idea that we observe the outside world by having it literally impress itself upon the surface of the eye. But it is also a picture of the world that we discussed in the previous chapter, a picture that is fundamentally unable to accommodate the Copernican revolution, and the triumph of Galileo over Aristotle. It is in short a very bad picture of the world.

The return of the magician

One helpful way to think about the scientific method then is as an attempt to distinguish causation from correlation. It is easy enough to spot patterns in the world around us. That is more or less what our brains are designed to do. But not all patterns are created equal, and only some of them indicate any kind of underlying mechanism that we can actually use to make predictions or design machines or cure diseases or do any other of the many things that we suppose it is the business of science to help us achieve. The problem however is that distinguishing between the genuine causal relations and the accidental correlations is often *hard work*. It requires lots of serious thinking, and lots of delicate experimental tests. It can require developing whole new mathematical techniques, interviewing untrustworthy sources, and sometimes even the need to stick a knitting needle into your own eye socket, and to wiggle it around violently in order to see what happens. One of the fundamental selling points of the Big Data revolution therefore was that it promised a way to finesse all of these difficulties. It offered a way to make a virtue out of mere correlation, on the assumption that if we only have *enough data*, if we only had the *biggest correlations*, then we would get the causally relevant relationships for free. Unfortunately, however, life is rarely that simple. In nature as much as in politics, there is no such thing as a free lunch.

Of course, that is not to say that the new generation of industrial number crunchers have nothing to add to the scientific enterprise. The fact that many of the most prominent examples of the Big Data revolution have turned out to be nothing more than an exercise in statistical naivety does not mean that all such approaches are doomed to fail. But there may be a deeper issue involved. Arguably, the distinction between causation and correlation is one of the *definitive* features of scientific practice. It is for example one of the most illuminating ways to understand how scientific thinking first arose from—and the way in which it can be contrasted with—the broadly magical thinking which it eventually replaced. In very crude terms, magical thinking tends to operate at the level of correlation. We observe superficial similarities between different objects or events, and conclude that they must therefore be linked. In the case of *sympathetic magic*, this might be because the two objects resemble one another in appearance. This is the sort of magical thinking that supposes that certain herbs can cure diseases because they look like the afflicted organ, or that you can injure an enemy by damaging a wooden doll carved in their likeness. In the case of *homeopathic magic*, the relevant similarity lies in the fact that one object has previously been in contact with another. This is the sort of magical thinking that supposes that you can cure a wound by treating the sword that inflicted it, bewitch the object of your affection by casting a spell over their nail clippings, or in the modern age, that a tiny vial of water can have medicinal properties because it was once diluted with a now absent active ingredient. In both cases, these magical connections only concern the surface properties of the objects in question—they rely upon a superficial correlation between the shape of a herb and the shape of an organ, or a correlation between the spatial location of various odds and ends. Scientific thinking by contrast is exemplified by the attempt to look beyond these correlations, and to see if any of these proposed connections demonstrate any kind of counterfactual stability.[7]

But magical thinking can sometimes be difficult to shake. We will recall that for Aristotle, all motion was the result of objects seeking to return to their natural place of rest—heavy objects towards the center of the universe, fire upwards towards the celestial spheres, and so on and so forth. Yet while we should not underestimate the empirical underpinnings of this account, nor disparage Aristotle's pioneering work in the scientific method, it must be conceded that the precise *mechanism* by which this system was governed was still essentially magical in nature. Different objects gravitated towards different places in the universe, not because of some attractive force or residual forward momentum, but rather an underlying similarity between the various components of the natural world, and some kind of residual connection to their place of origin. Fire goes up and earth goes down because that is where they are *supposed to be*, not because of any external force.

And one of the reasons why such thinking can be so difficult to shake, and one of the reasons why Galileo had such difficulty in overthrowing the Aristotelian picture, is because magical thinking often stretches beyond explaining the events of the physical world into providing an explanation for the moral or spiritual world as well. At its height, the Aristotelian picture of the world—reinterpreted and reimagined by the medieval church—presented a seamless account of almost every aspect of man's existence. Heavy objects fall towards the center of the universe, and insofar as man succumbs to his material impulses, so too is he drawn downwards, and onwards to Hell itself, conveniently located at the most central point of all. But man is also a being of spirit, which like fire is drawn upwards to the crystalline spheres, and as it becomes ever more rarefied, towards God in his Heaven beyond the outermost stars. It was a physics and theology all in one, and there was a very real concern that with the abandonment of Aristotle's principles of motion, so too would follow moral anarchy and existential angst. When Copernicus and Galileo were attacked for displacing man from the center of the universe, the charge was not simply concerned with the spatial coordinates of the Earth in relation to the rest of the Solar System. The scientific revolution overthrew man's moral position too, casting him adrift in a universe no longer structured to provide him with spiritual guidance. It is of little surprise therefore that Leibniz was able to accuse Newton of being unable to find a place for God within his mechanistic world view, or that Newton took the accusation so seriously.

We should not, however, be too quick to scoff at such concerns. The willingness with which some have embraced the Big Data revolution without apparent regard for its statistical underpinnings has more than a little tinge of magic about it. After all, it is the explicit privileging of correlation over causation—backed up with the powerful talismanic devices of having come out of a computer, widely acknowledged as the great prophets of our age. It also parallels a similar pattern in our political and moral thinking, which also tends to emphasize the *symptoms* of some underlying injustice, rather than the injustice itself. You only have to spend a few minutes on social media to witness the almost obsessive attempt to police politically incorrect language, which is often seen as a legitimate substitute for actually battling prejudice or engaging in any wider social action. Such exchanges are hilariously funny, since they frequently degenerate into an exercise of moral oneupmanship amongst socioeconomic groups who have never faced any prejudice themselves (I believe this is called "checking one's privilege"), and because the idea that it is words *themselves* that have power or moral significance—quite independently of the speaker or the speaker's intentions—is itself one of the most paradigmatic examples of magical thinking available.

It is of course difficult to know whether or not it is a general decline in moral thinking that has encouraged a more superficial approach to scientific investigation, or an unwitting error in scientific investigation that has encouraged this kind of moral laxity. Or perhaps both are merely a symptom of a third, underlying cause, such as mankind's general reluctance towards hard work whenever an easier option presents itself. Perhaps the truth is that we are just *lazy*. But then again, maybe there's no causal relationship between the two at all, and it's just another instance of an unconnected correlation assumed to hold greater significance than it actually deserves.

4

88.6 percent of all statistics are made up

The crackling fire threw flickering shadows across the wood-paneled sitting room. Outside, the early evening gloom gathered against the windowpanes, and the snow-muffled rattle of carriages drifted up from the streets. Leaning back into his voluminous armchair and pressing the tips of his fingers together, he turned to his friend and spoke in a slow and measured voice:

> "It is simplicity itself," said he; "my eyes tell me that on the inside of your left shoe, just where the firelight strikes it, the leather is scored by six almost parallel cuts. Obviously they have been caused by someone who has very carelessly scraped around the edges of the sole in order to remove crusted mud from it. Hence, you see, my double deduction that you had been out in vile-weather, and that you had a particularly malignant boot-slitting specimen of the London slavery. As to your practice, if a gentleman walks into my rooms smelling of iodoform, with a black mark of nitrate of silver upon his right forefinger, and a bulge on the side of his top hat to show where he has secreted his stethoscope, I must be dull indeed if I do not pronounce him to be an active member of the medical profession."[1]

The speaker is of course none other than Sherlock Holmes, the World's Greatest Detective and high-functioning cocaine-addict. Doctor Watson is predictably amazed by the almost supernatural acumen of his friend, and readily confirms that he has been out in the rain, that his servant girl has been remiss when it comes to cleaning his boots, and that he has indeed recently resumed his medical practice. And with that, in walks a poorly disguised member of the Bohemian Royal Family, and so begins another adventure of missing diamonds, mistaken identities, and murder amongst the upper classes.

Leaving to one side some of its more fanciful applications, there is clearly something to Holmes' practice, and it has often served as a source of inspiration for thinking about the scientific method. There is, however, one important caveat in order. It is rather unfortunate that the Great Detective insists upon referring to his style of inference as *deduction*, and that he repeatedly informs Watson that he has merely *deduced* the solution from the available facts. It is nothing of the kind. Technically speaking, a deduction is a purely logical procedure, of the kind that we might use in mathematics or something we could program a computer to do. It is the sort of inference where, if we know that Socrates is a man, and we know that all men are mortal, we can indeed *deduce* that Socrates is not going to live forever. Given the premises, there is absolutely no way for the conclusion to be false. It is the sort of certainty we have in knowing that if someone is a bachelor, they must also be unmarried, or that two plus three equals five. But for all his ingenuity and brilliance, this is clearly not the sort of inference that Sherlock Holmes so often performs—the actions of a careless servant girl may well explain the scuffs on Watson's boots, but this is certainly not the *only* explanation for the observed damage.

Indeed, I had always liked to imagine that somewhere in Arthur Conan Doyle's papers there exists a series of outtakes from *The Adventures of Sherlock Holmes*, a collection of deleted scenes where Holmes waxes forth on his so-called deductive method, only for Watson to point out the absurdity of it all and knock his insufferable colleague down a peg or two. "What utter nonsense, Holmes," he would exclaim. "I scraped my boot climbing over a fence the other day. I wouldn't usually have undertaken such foolish activity, but the weather was so delightful I couldn't resist my boyish urge. I have quit the medical profession altogether you know, and taken up a new career as a mushroom collector. I was out mushroom collecting only this morning in fact, and came across a fantastic growth of Angel's Bonnets—which as I am sure you are aware, Holmes, are a natural source of iodoform and unmistakable in their odour. Not wanting to spoil the lining of my jacket, I naturally enough stashed them under my hat, thus accounting for both the persisting smell and the unusual bulge. And the stain on my finger is nothing more than ink from a cashier's cheque, freshly received in payment of some of the mushrooms I sold this afternoon to a fashionable restaurant."

Or maybe Watson simply looks at Holmes sadly and shakes his head. "I'm afraid not, old boy. I haven't practiced medicine for years. But I had to share a train carriage this morning with a fellow who absolutely reeked of disinfectant. A hospital orderly, I'd wager. It really was most inconsiderate for the other passengers, and I took it upon myself to admonish the scoundrel. We came to blows, and while I ended up giving him a damn good thrashing—I was in the army Holmes, as you remember—I managed to scrape my shoes, and my top hat took quite a beating. The unsightly bulge is the result of my trying to bash it back into shape. By the time we reached Waterloo, the police had been summoned, and I was forced to sign a statement explaining how it was all the other fellow's fault. As you can see, I still have some ink on my fingers, and that blasted smell has followed me all the way from the Home Counties."

Another way to put the point is to note Holmes' famous dictum that once you eliminate the impossible, whatever remains—no matter how improbable—must be the truth. This does indeed provide us with a straightforward recipe for performing a deductive inference. If you really have eliminated *all of the other alternatives* except for one, then your conclusion really can be logically unassailable. The problem is that this is not something that the Great Detective ever actually does—and for good reason too, since for any particular fact that you might want to explain, the number of other alternatives that you would need to eliminate before you get to the truth would be simply too vast to contemplate. Picking mushrooms and getting into a fight on the train are only the tip of the iceberg.

So whatever it is that Sherlock Holmes does, it is not *deduction*. And this is in fact just as well, since we have already seen just how badly deduction fares as an account of the scientific method in Chapter 1, when we discussed the work of Karl Popper. The whole idea of falsificationism is after all itself an instance of eliminating the impossible in order to end up with the truth. As we have already seen though, falsifying a scientific theory is not a straightforward affair. No matter how conclusively our theory may seem to fail a test, it will always be possible to find some other factor on which to lay the blame. If we were testing a theory of planetary motion for example, any discrepancy we have with our observational data could always be due to faulty equipment, careless assistants, or some unexpected interference we had failed to take into account. That is to say, we don't just need to eliminate one wayward theory of planetary motion—we also need to eliminate any doubts we may have regarding the accuracy of our measuring equipment, the existence of rogue asteroids, errors made by a lazy graduate student, catastrophic extraterrestrial volcanic activity, previously unnoticed additional planets ... and so on and so forth. Thus, just as Sherlock Holmes never really eliminates all of the impossibilities before drawing his conclusion, neither does the practicing scientist proceed on the basis of simply trying to show that his theories are false.

But it should also be clear that Holmes' procedure does not neatly correspond to any other account of the scientific method that we have so far considered. It is not merely an instance of making a pure, disinterested observation of the data—which is also just as well, since we have seen that this does not provide a coherent understanding of scientific practice either. Indeed, throughout his adventures, Holmes frequently notes how he has observed precisely the same things as Watson, yet has managed to draw inferences that have simply eluded the good doctor. The fact is of course that all observation requires an interpretation. Ptolemy saw the Sun orbiting the Earth, while Copernicus saw the Earth orbiting the Sun. Aristotle saw bodies moving towards their natural place of rest, while Galileo saw bodies undergoing inertial motion. And while Watson saw only a scratch on the soles of his boots, Holmes saw the workings of a lazy and disgruntled employee.

Nor is Holmes' method a simple case of induction. He does not proceed from a set of regularities that have been observed in the past, to an inference about how they will continue in the future. Almost all of the inferences that Holmes makes are unique. He does not have a journal recording previous instances of scuffed boots, alongside a list of exhaustive investigations into their cause. He does not think to himself that in 80 percent of the previous scuffed soles I have investigated, the damage was due to a careless servant girl, and so conclude on the strength of probability that Watson needs to fire his employee. And nor does Holmes proceed on the basis of carefully controlled experiments. At no point in *The Scandal in Bohemia* does he give Watson a number of identical pairs of boots and ask him to wear them in different situations and on different days of the week in order to isolate a common factor in all resulting wear and tear. And even if the option had been available, he definitely wouldn't just google the result.

A scandal near Bohemia

Just what then is Holmes doing when he makes his astonishing discoveries? The answer turns out to be far from elementary, but fortunately we can begin to make sense of the process with the help of our very own Doctor Watson. This is the story of Ignaz

FIGURE 4.1 *"Bollocks," replied Watson.*

Semmelweis, a Hungarian-born physician who practiced at the Vienna General Hospital in the late 1840s. Semmelweis is famous for being one of the first individuals to suggest that maybe doctors should wash their hands before they examined a patient. It was a pretty outrageous suggestion at the time, since not only did it contradict received medical wisdom by suggesting that diseases could be transmitted by human contact, it was also *damned impertinent* as it implied that doctors—men of breeding and expensive educations, and in every other way your social superiors, so watch your manners—might be anything less than pristine and fragrant. It also required some serious logical reasoning, which is why philosophers of science love talking about Semmelweis *almost as much* as they love talking about black ravens.

When Semmelweis first arrived at the Vienna General Hospital, one of his main responsibilities was to oversee the First Obstetrical Clinic, a maternity ward that offered free treatment for vulnerable and disadvantaged women—I believe the modern terminology is "prostitute"—in exchange for providing valuable hands-on experience for medical students. The Second Obstetrical Clinic next door offered similar treatment, only in this case for the benefit of trainee midwives. The problem facing Semmelweis was a radical disparity in mortality rates between the two clinics. Childbirth was always something of a risky business in the nineteenth century, but even by those standards the First Obstetrical Clinic was an absolute deathtrap. Roughly 10 percent of all mothers who gave birth in the First Clinic died of childbed fever shortly afterwards, in contrast to around 4 percent of mothers in the Second Clinic. Semmelweis records extraordinary scenes of heavily pregnant women begging to be admitted to the Second Clinic—admission tended to alternate by days of the week—and many preferred to give birth on the streets than face the alternative.

Understandably horrified by the state of affairs, Semmelweis immediately began a systematic investigation into why the First Obstetrical Clinic was so dangerous. He quickly dismissed two of the most popular explanations that had been offered for the poor performance of his clinic. The first maintained that the First Clinic was badly overcrowded, and that this had an unhealthy impact upon the safety of its patients. The problem with this explanation was that it was just

88.6 PERCENT OF ALL STATISTICS ARE MADE UP

straightforwardly false—the Second Clinic was in fact considerably more crowded than the First, precisely because everyone knew that it was so much safer (recall all those desperate mothers giving birth on the streets rather than take the risk). The second proposal was that the Vienna General Hospital was under the influence of a miasma of unhealthy vapors—a sort of fog of disease and vice and general disreputable living that provided both an early pre-cursor to modern germ theory and a physical embodiment of the bourgeois disdain for the working classes all in one pseudoscientific package. The problem with this though was that even if it was true, it couldn't provide any explanation for why the First Clinic performed so much *worse* than the Second Clinic.

Having thereby exhausted current medical expertise on the matter, Semmelweis was forced to start making his own hypotheses. One obvious difference between the two clinics was of course that one had medical students, while the other had trainee midwives. Semmelweis conjectured therefore that perhaps the medical students performed rougher examinations than the midwives, since they would have all been men and presumably less careful and experienced with handling female patients. If anything however, many of the midwives were considerably more brusk and burly than the largely bookish medical students. Another difference was that in the First Clinic women delivered lying on their backs, while in the Second Clinic they delivered lying on their sides. Semmelweis couldn't think of a particularly good reason why this should make a difference, but he nevertheless instructed that all deliveries should take place in the same position across both clinics—a good example of controlling a variable—but unfortunately this had no effect on the mortality rate.

More imaginatively, Semmelweis noted that there was a difference in the layout of the two clinics, and their location with respect to the rest of the hospital. More specifically, he noticed that if a priest came to the hospital in order to deliver the last rites to a patient in the upper wards, he had to walk straight through the middle of the First Clinic, while avoiding the Second Clinic altogether. Semmelweis therefore even considered the possibility that the sight of the solemn-faced priest stalking the corridors filled his patients with an overpowering sense of dread, and it was this that made them more susceptible to illness and infection. The priest was duly instructed to vary his route while Semmelweis carefully plotted the ongoing deaths in his clinic. But again, none of this seemed to make any difference at all.

Semmelweis' breakthrough finally came after one of his colleagues accidentally cut himself while performing an autopsy and died of an illness with similar symptoms to the childbed fever ravaging the First Obstetrical Clinic. Crucially, Semmelweis noted that while the medical students working in the First Clinic also performed autopsies as part of their rotations—often immediately before attending to the expectant mothers—this was obviously not part of the training for the midwives in the Second Clinic. Semmelweis conjectured therefore that many of the medical students in the First Clinic transmitted "cadaveric matter" from the corpses in the morgue to the mothers in the clinic, and it was this that was ultimately responsible for the lethal infection. A policy of rigorous disinfectant was imposed for the students, and the mortality rates for the First Clinic dropped significantly, even lower than those for the Second Clinic.[2]

The important point to note in all of this is that Semmelweis' reasoning involved neither inductive extrapolation or deductive elimination. He couldn't simply work his way through all the possible reasons for the rampant childbed fever in the First Clinic until only one remained, since the number of alternatives was far too vast. And nor was he in a position to observe many different obstetrical clinics with many different medical procedures and feed it all into an algorithm. In many ways, in fact, Semmelweis seems to have the whole scientific method back-to-front. It is natural to think of an explanation as being one of the *consequences* of a predictively successful scientific theory. We assemble our data, construct a theory on the basis of that data—perhaps by extrapolation from that data, or eliminating those theories that cannot accommodate it—and then offer an explanation in terms of the central concepts employed by the resulting theory. But rather than constructing a scientific theory, and then using it to explain the evidence before him, Semmelweis began by formulating an explanation, and then using it to construct his theory. If cadaveric matter transmitted from the morgue to the clinic was a significant source of infection, that would provide a good explanation for why medical students were a higher risk than midwives—just as if Watson did have a careless servant girl, that would explain why he has scuffs on his boots. It is only once we have a good idea of what the explanation might be, that we have any idea what the relevant evidence is. For Semmelweis and for Holmes, a satisfying explanation is what *generates* a good scientific theory, not the other way around.

This style of reasoning is, in fact, implicit in much scientific practice, although it is not always clearly distinguished from the more straightforward instances of induction discussed in the previous chapter. The archaic name for this method of reasoning is *abduction*, although nowadays it goes by the more user-friendly name of *inference to the best explanation*, which is about as literal a description as you can get. Quite simply, the idea is that we should infer the truth of the scientific theory that provides the best explanation for the evidence available—whether or not that theory offers the only possible explanation available, and irrespective of whether that theory provides the most natural way of extrapolating from that evidence. To take a simple example, both Ptolemaic and Copernican astronomy allow us to predict the future location of a planet with more-or-less the same degree of accuracy, yet they offer radically different explanations as to why the planet in question will be where they say it will. According to the Ptolemaic account, this is because the planet follows a (complex) series of epicycles centered around the Earth; whereas according to the Copernican account, this is because the planet follows an (equally complex) series of epicycles centered around the Sun. In much the same way, Semmelweis considered a number of different theories that—at least initially—all seemed to predict the different rates of childbed fever in his maternity wards, although the explanations offered ranged across everything from clumsy medical students and religious anxiety, to birthing positions, epidemic influences, and the presence of cadaveric matter. In broader terms, the idea is that we need to acknowledge that good scientific practice is often very difficult to shoe-horn into any particular set of rules or regulations, and that at the end of the day, we need to allow good judgment, intuition, and general considerations of overall plausibility to help guide our investigations.

To take just one more example, the wide-spread scientific acceptance of the principles of natural selection are clearly an instance of inference to the best explanation. The available evidence—the existence of biological complexity and its varying degrees of environmental adaptability—certainly does not logically entail the contemporary biological world view. After all, the very starting point for much of our discussion so far has been motivated by the fact that there are any number of alternative proposals that are logically compatible with that evidence. Indeed, in some of the more recent incarnations of creationism, these theories have been intelligently designed so as to accommodate as much of the contemporary evidence as possible. By the same token however, it is also clear that scientific confidence in evolution by natural selection has very little to do with the traditional inductive method of extrapolating from the evidence. There is after all only one natural world to consider, which makes the possibility of drawing any general conclusions somewhat difficult. Similarly, we can hardly conduct any carefully controlled experiments concerning the initial conditions of life. The reasoning rather is that gradual evolution through genetic mutation and environmental pressure just offers a *better explanation* for biological complexity than the work of an all-powerful creator. Of course, many creationists may well disagree as to whether evolution really does offer the better explanation; the point however remains that this appears to be the form of reasoning employed.

A scientific explanation for the success of science

It is always tempting to try and reduce the scientific method to a precise set of rules or algorithms. If we could only come up with the exact recipe for good scientific practice, then we would be able to apply the same tools and techniques for the benefit of other, less successful fields of study. The entire process of investigation and experimentation could finally be systematized and standardized—you just follow the instructions on the packet, and sooner or later the secrets of nature will be revealed to you. The truth, however, is that good scientific practice is often more like a balancing act. Sometimes we allow our best explanations to guide our choice of theory, rather than allowing our choice of theory to determine our available explanations. It is a matter of give and take, and often depends more upon the subtlety and insight of the scientist in question than it does upon any predetermined set of principles.

It is one thing to acknowledge the Holmesian quality to much scientific investigation, and the importance played by inference to the best explanation in our scientific methodology. It is, however, quite another thing to determine whether or not this is a good thing. We might reasonably wonder if inference to the best explanation really does provide a reliable guide to our choice of theory. After all, while it undoubtedly worked in the case of Semmelweis, it also seems simply to reintroduce all of those elements of human foible—our guesses and intuitions, and not to mention the vague and unquantifiable notion of one explanation being *better* than another— that we hoped that the rigorous application of the scientific method would eliminate. As it turns out, however, the idea that our scientific methodology is ultimately based upon a process of abduction or inference to the best explanation has in fact been used to provide a very famous argument for the reliability of our scientific theories, and for the truth of science in general.

While you will be able to find examples of this line of thought in many different places, the earliest modern articulation of the argument was offered by the Australian philosopher J. J. C. Smart.[3] It is at heart a kind of *plausibility* argument. It begins with the observation that our scientific theories have proved to be extremely successful in a wide range of applications, from the predictions they make about the future, to the technological advances they support. We rely upon them every time we step onto an aeroplane, or turn on our computers, or take the medicine prescribed to us by a doctor. To put it in a nutshell, *science works*. It seems reasonable then to suppose that our scientific theories must be more or less true, that protons and electrons really exist, and that scientists are indeed generally reliable when it comes to investigating the external world. The only alternative would be to put the whole thing down to chance. We would have to imagine that everything we know about physics is completely false—but that somehow when we come to build an aeroplane or assemble a computer, all of these mistakes manage to cancel each other out, and that every minute of every day we avoid a certain and painful death by nothing more than persistent dumb luck. On this view, the success of science would be what Smart calls a "cosmic coincidence," a possibility he dismisses with forthright antipodean disdain.

But we can perhaps even go one further. Human beings are after all just another part of the natural world, and our cognitive processes—our beliefs and our desires, and how we choose to act upon them—are as legitimate a field of scientific investigation as any other naturally occurring phenomenon. In particular then, the scientific theories that we construct in order to explain the world around us can *themselves* be the object of further scientific study. In much the same way that a zoologist might study the ways in which different animals respond to their environments, or an anthropologist might study the primitive tools of our distant evolutionary ancestors, so too can a cognitive psychologist study our belief-forming mechanisms and investigate whether or not they are likely to be successful.

The idea then is that when we come to ask about the reliability of our scientific theories, this is in itself a broadly scientific question. This then allows us to put an interesting spin on Smart's original argument. For not only can we argue that it is just *philosophically* more plausible to suppose that our scientific theories are approximately true, we can in fact argue that it is *scientifically* more plausible to suppose that our theories are approximately true. This is because the truth of our scientific theories is the *best explanation* for their predictive success—and as we have seen, it is generally considered to be good scientific practice to infer the truth of our best explanations. Developing upon Smart's original proposal, the American philosopher Hilary Putnam argued that assuming our scientific theories to be approximately true:

> is the only philosophy that does not make the success of science a miracle. That terms in mature scientific theories typically refer ... that the theories accepted in a mature science are typically approximately true, that the same terms can refer to the same even when they occur in different theories—these statements are viewed not as necessary truths but as part of the only scientific explanation of the success of science, and hence as part of any adequate description of science and its relations to its objects.[4]

The way in which we theorize about the natural world is a scientific phenomenon, and according to our best scientific methods, we have good reasons to believe that this is a reliable process.

Picking up on Putnam's phrase, this line of reasoning has become known as the *No Miracles Argument* in the contemporary literature. Many philosophers of science—the present author included—have spent a considerable portion of their professional careers trying to decide whether or not it is a good argument. But before doing so, it is important to be clear on what exactly the argument attempts to achieve. It is not an argument intended to convince the entrenched skeptic that they should in fact believe whatever science has to tell them. After all, there is a sense in which the argument is a bit circular. It is an argument that appeals to our scientific methods in order to conclude that our scientific methods are actually reliable. And if you don't already accept that science is more or less in the business of delivering reliable knowledge about the world around us, you will not be terribly convinced to be told that science itself tells us that science is reliable. That would be like suddenly believing a particularly untrustworthy politician just because he promised us that this time he was telling the truth.

Rather, the argument attempts to show us that believing our scientific theories to be generally reliable is part of a coherent world view. The fact that we have good scientific reasons to trust our scientific theories is not a foregone conclusion. Some methods of investigating the world can actually be self-undermining. Consider, for example, someone who tells us that all psychics are frauds, and that he knows this because his local tarot-reader saw it in the cards. Or someone who thinks that we shouldn't attempt to make an inductive inference on the basis of our current evidence, because we know full well that all our previous attempts to extrapolate from the past have been unsuccessful. Or to take an even simpler example, imagine our untrustworthy politician agreeing with us and announcing that all politicians are liars.

We may never be able to prove beyond all reasonable doubt that our scientific theories are true. The world is a complicated place, and man is a fallible creature. But what we can do is to try and show that our beliefs are all part of an overall rational package. One of the most important mechanisms involved in our scientific methodology is the practice of inferring the truth of our best explanations, and using them as a guide for constructing our scientific theories. And the reason why we should suppose that inference to the best explanation is a reliable method of reasoning is precisely because it is part of that coherent world view.

An evolutionary alternative

What the foregoing reasoning tries to establish is that the details of our scientific methodology, and our critical evaluation of those methods, can in fact positively reinforce one another. It is a sort of epistemological feedback mechanism. We begin by arguing that the approximate truth of our scientific theories is the best explanation for their predictive success, and that therefore we have good reasons to believe that our scientific theories are indeed generally reliable. However, realizing that our argument is an instance of inference to the best explanation, we might then reasonably ask why it is that we should rely upon such a method of reasoning. Upon further investigation, we note that inference to the best explanation is a widespread feature of our day-to-day scientific practice—a practice that we already believe produces scientific theories that are approximately true. So our scientific methods must be reliable, which means in particular that inference to the best explanation must be reliable. And if inference to the best explanation is reliable, then our original conclusions are confirmed, and off we go around the philosophical roundabout one more time.

It is certainly an intellectually pleasing exercise. But just because a world view is internally consistent, it does not necessarily mean that it is correct. There may be other consistent world views that we can adopt. Perhaps Watson has recently resumed his medical practice, stashed his stethoscope in his top hat, and employed a careless servant girl. But perhaps he has also packed it all in for the unparalleled excitements of mushroom hunting. It might well be the case that gradual evolution through natural selection is the best explanation for the existence of biological complexity—and that such a belief in turn reassures as to the reliability of the scientific methods upon which the theory is based. But similarly, a committed creationist might argue that divine providence provides the best explanation for the existence of biological complexity—and since his methodological principles also entail that he should treat the Book of Genesis as a literal account of the origins of the world, he will also be able to boast an internally consistent world view that positively supports his reasoning.

But perhaps more importantly, the No Miracles Argument as presented above is simply far too abstract to have any real purchase. It is one thing to argue that scientific practice involves inferring the truth of our best explanations, and that in turn, the truth of our scientific theories is the best explanation for their success. But without a more concrete understanding of how one explanation can be better than another, and indeed what sorts of explanations are actually treated as being better than others within our most successful scientific practices, we are in no position to prefer the evolutionist over the creationist, or have any more reason to suppose that Watson practices medicine than that he spends his time collecting rare mushrooms.

So what would a more concrete scientific explanation for the success of science look like? According to the philosopher Bas van Fraassen:

> The success of science is not a miracle. It is not even surprising to the scientific (Darwinist) mind. For any scientific theory is born into a life of fierce competition, a jungle red in tooth and claw. Only the successful theories survive—the ones which *in fact* have latched on to actual regularities in nature.[5]

The idea is that if we really are serious about relying upon our best scientific theories as a guide for our more philosophical speculations, then we should look at the details of those scientific theories. In particular, it has been one of the great intellectual achievements of the modern scientific era to realize that well-ordered systems can be explained as the result of random chance operating within a competitive environment—and not necessarily in terms of an all-powerful agent overseeing that everything fits together in the right sort of way. It follows then that in order to explain the success of our scientific theories, we needn't suppose that these theories are true, or that our scientific methods are particularly reliable. We only need to suppose that there is some mechanism in place whereby the *unsuccessful* theories are eliminated.

Smart argued that if our scientific theories were not approximately true, then it would be absolutely amazing that we haven't all died a grisly death by now. But that oversimplifies the situation. There are lots of false scientific theories that nevertheless work extremely well. Technically speaking, Newtonian Mechanics is false, which is why it was superseded by Einstein's theory of relativity. But within a certain range of applications—for systems traveling at a speed significantly slower than the speed of light—its predictions are close enough to the truth for us not to notice the difference. The trick of course is finding a false scientific theory that nevertheless works within the relevant range of applications. But suppose now that we begin with a large number of different scientific theories, all making competing claims about the world. As time goes on, some of these predictions will be vindicated, and others shown to be wrong. Whenever this happens, we eliminate the unsuccessful theories from the pool. We might say that they have proven to be poorly adapted to their environment and have perished. At any particular moment of time, however, all of the scientific theories currently under consideration will be predictively successful—not because they are true nor because our scientific methods are reliable, but merely because we have *selected* those theories that are predictively successful.

Indeed, once we start considering the content of our favorite scientific explanations, further considerations come to light. Cognitive psychology, for example, shows us that human reasoning often relies upon a series of heuristics designed in order to help simplify complex calculations.[6] In the case of perceptions for instance, the judgments we make regarding how far away an object is are often based upon considering how clearly we can observe it. The sharper our image of the object, the closer it must be. This works well as a rough and ready assumption in a large number of cases, but it can also lead us astray—which is why we tend to overestimate distances when the visibility is poor, and why car wing mirrors come with a helpful warning.

88.6 PERCENT OF ALL STATISTICS ARE MADE UP

A similar situation holds when we are making judgments under uncertainty. When Sherlock Holmes infers that his friend Doctor Watson has resumed his medical practice, he is making a probability judgment. More specifically, he is making a probability judgment based upon some of the salient features of his friend's appearance, such as the smell of hospital disinfectant, the silver nitrate stains on his fingers, and the unusual bulge in his top hat. These are all characteristics that we might naturally associate with being a doctor in the late nineteenth century—along with perhaps coming from an affluent background, going to the right school, and maybe even having a penchant for an elaborate moustache. In other words, Holmes is making a probability judgment on the basis of how *representative* Watson is of the various doctors of the period. Allowing the representativeness of a sample to guide our probability judgments is a very common cognitive heuristic. If a very high proportion of doctors smell of hospital disinfectant, then the fact that this man also smells of hospital disinfectant should make it very likely that he too is a doctor. It is the sort of pattern recognition that the human brain enjoys, a way of comparing like with like, and one that offers a quick and ready rule of thumb for estimating difficult probabilities. The problem of course is that like any other kind of heuristic, this one can also be extremely unreliable.

One particularly clear example of how a reliance upon representativeness can lead us astray occurs in games of chance. We all know that if we toss an unbiased coin enough times, the number of heads will equal the number of tails. It does not follow however that for any short run of tossed coins, the number of heads will exactly equal the number of tails. There is no reason to suppose that if you only toss the coin ten times, for example, that we will get an equal number of heads and tails. Yet nevertheless, we instinctively feel that it must be *less likely* for us to get an unbroken run of consecutive heads than for us to get an even mix of results, and the reason for that is because an even mix of heads and tails strikes us as *more representative* of the sequence as a whole. Yet it is precisely this kind of reasoning that leads to the well-known gambler's fallacy—the more times we see the roulette wheel comes up on red, the more likely we think that the next spin will come up on black, and the more money we end up losing at the casino.

Another example, and one that is more immediately relevant to our present concerns, is the way in which we often allow our judgments of representativeness to override other probabilistic considerations. It may be well known that a far larger proportion of today's adult population work in an office than outside on a farm. Yet if we are told that Charlie is a big, strapping lad with a weathered face and strong, gnarled hands, we instinctively feel that it is more likely that Charlie makes his living off the land than it is that he spends his days pushing a pencil across a desk. After all, working on a farm requires some serious elbow-grease, whereas sitting behind a desk all day only tends to bugger up your back. Let us suppose then for the sake of argument that only a tiny fraction of office workers have weather-beaten faces and rugged hands—perhaps they are devoted gardeners in their spare-time, or outdoor-sports enthusiasts. The problem, however, is that if the proportion of office-workers to farmers is large enough, this tiny fraction might still be larger than the total number of farmers. And if that is the case, it is still more likely that Charlie operates a desktop rather than a tractor, regardless of how well he represents our stereotype of a farmer.

This sort of error is known as the *base-rate fallacy*, since it is an attempt to make a probabilistic judgment that nevertheless ignores the prior probability—or base rate—for the case in question. In the 1950s for example, the German-American psychiatrist Fredric Wertham argued that comic-books had a negative influence on adolescent minds, citing the fact that the vast majority of troubled teenagers treated by his clinic were devoted comic-book fans. His best-selling *Seduction of the Innocent* bolstered this analysis by denouncing Superman as a fascist, and making snide insinuations about Batman's relationship with Robin. Nevertheless, Wertham's correlations were considered striking and a matter for urgent alarm—until of course it was realized that at the time, roughly 90 percent of *all teenagers* in the U.S. were devoted comic-book fans. Wertham's argument was thus a classic instance of the base-rate fallacy, as indeed have been more recent scares about the effects of television, rock music, and even all those action movies from the 1980s where Jean-Claude Van Damme does the splits and punches a bad guy in the groin, and which definitely did not have any kind of influence upon me as a child whatsoever.

88.6 PERCENT OF ALL STATISTICS ARE MADE UP

Now let us consider again the No Miracles Argument encountered above. This maintains that the best explanation for the success of science is that our theories are more or less true, and therefore that our methods are generally reliable. We can see now that one of the intuitions lying behind this argument is a consideration of representativeness. For on the one hand, it is of course extremely likely that any scientific theory that is true is also one that is predictively successful. On the other hand, it is extremely unlikely that any scientific theory that is false should also manage to make any accurate predictions about the world. We therefore instinctively feel that it is more likely for a predictively successful scientific theory to be true. But that would be to just commit the base-rate fallacy all over again. It all depends upon *how many* scientific theories we have to consider, and how likely it is for any one of them to be true.

As van Fraassen noted above, if we start off with enough different scientific theories to consider, and systematically eliminate those that encounter difficulties, we should not be terribly surprised if those that we end up with are generally successful. We can express the same point in terms of the base-rate fallacy. Let us suppose that the probability of a true scientific theory being predictively successful is very high, whereas the probability of a false scientific theory being similarly successful is very small. But now let us also suppose that out of the entire range of scientific theories to consider, there are significantly many more false scientific theories than true ones. In such a situation, it might well be the case that the overall number of false scientific theories that are nevertheless successful is actually *greater* than the overall number of true scientific theories that are successful. And if that is the case, then it is in fact more likely for a predictively successful scientific theory to be *false* than it is for it to be true.

It all comes down to the overall sample of scientific theories under consideration. On the one hand, if most of these scientific theories are true, then the No Miracles Argument goes through. While on the other hand, if most of these scientific theories are false, then the No Miracles Argument fails. It follows then that in order to properly evaluate the No Miracles Argument we need to know something about the overall distribution of potential scientific theories. We need to know for any arbitrary scientific theory that might be in the pile, how likely it is for that theory to be true, and how likely it is for that theory to be false. And that is something we simply do not know. Indeed, that was the whole point of formulating the No Miracles Argument in the first place! It was supposed to be an argument for believing our scientific theories to be true—but now it turns out that in order for the argument to work, we *already* need to know whether or not our scientific theories are true. The whole approach, painstakingly tweaked and developed by philosophers of science for years across countless books and journal articles, has been a complete and utter waste of time.

Inference, explanation, and the miracle of science

When Sherlock Holmes makes his astonishing pronouncements about his friend's private life, he is in reality making a series of probability judgments about the matter in hand. More specifically, he is making a series of probability judgments based upon the representativeness of the sample in question. He concludes that Watson must have resumed his medical profession, since Watson smells of hospital disinfectant, and because the vast majority of doctors at the time similarly smelt of hospital disinfectant. In doing so, however, Holmes is relying upon a cognitive heuristic that runs a considerable risk of error. Given the relatively small proportion of doctors operating in Victorian London—at least as compared to all the other professions that could conceivably come into contact with iodoform in some way or another—the probability that Watson has returned to his medical practice, given that he smells of disinfectant, is actually rather low.

Of course, Holmes has a considerable advantage over the rest of us mere mortals. Not only is he a remorseless calculating machine of considerable private resources and a regular supply of good-quality cocaine. Most importantly, he is a fictional character, and his actions are guided by an omnipotent author who ensures that his cognitive speculations are always absolutely spot on. For the World's Greatest Detective, making a probability judgment on the basis of the representativeness of a sample is not a quick and dirty heuristic designed for saving time—it is in fact a highly reliable intellectual technique, finely calibrated so as to deliver unimpeachable knowledge of the external world. To put the same point another way, Sherlock Holmes can continue to make his "elementary" inferences with smug self-assurance because his literary creator has guaranteed that the base-rates will always be in his favor.

Sadly, life in the real world is not always so straightforward. We do not have the same assurances that our own cognitive faculties are as well calibrated with the world around us. The various examples discussed above are all testimony to this unfortunate state of affairs. But there is perhaps a deeper explanation for this fact. According to our best scientific theories, man has evolved from simpler organisms, and his cognitive faculties are the result of endless trial and error in the face of a hostile environment. Such selection pressures do not guarantee a set of intellectual tools designed to deliver reliable and accurate information about the world around us, but rather a set of intellectual tools designed to *keep us alive* in that world—and this does not always amount to the same thing. A primitive ancestor who concludes that every little movement in the corner of his eye is evidence of a tiger, and is forever running for his life, may in fact have a much greater chance of survival than his more discerning colleague, since it only takes one small mistake in order to permanently eliminate such an individual from the gene pool.[7]

So, evolution may well select for individuals who avoid danger, rather than those who can correctly identify it. Another problem is that no matter how reliable our cognitive faculties may be, they have all evolved within an incredibly limited environment and under a very limited set of conditions. We might then reasonably wonder whether or not such a set of intellectual tools is still relevant for today's concerns. Charles Darwin himself raised just such a concern. Writing to his friend William Graham in 1881, he admits that:

> with me the horrid doubt always arises whether the convictions of man's mind, which has been developed from the mind of the lower animals, are of any value or at all trustworthy. Would anyone trust in the convictions of a monkey's mind, if there are any convictions in such a mind?[8]

Our early ancestors spent much of their time hunting large animals across the Serengeti. It might then be reasonable to suppose that our cognitive faculties are reasonably reliable at spotting medium-sized objects moving at slow speeds across a flat and sunny plain. Such expertise, however, does not easily translate to the domains of modern science, when objects can move so fast that they experience time dilation, or when objects can be so small that they simultaneously exist as both particles and waves. And that is only the shallow end of contemporary physics. Why should we suppose that such intellectual capacities have any value whatsoever?

All of this puts a very different spin on the No Miracles Argument. According to philosophers like J. J. C. Smart and Hilary Putnam, we have good reasons to believe that our scientific theories are largely true—and that therefore the scientific methods we use to generate them must be generally reliable—since if our scientific theories *weren't* approximately true, it would be simply miraculous for them to be as successful as they are. But once we begin to reflect upon the evolutionary history of our cognitive faculties, and all the various ways in which we know that they can go wrong, it can be tempting to turn the whole argument on its head. Forget about predictive success: the real miracle would be if our scientific theories were in any way true at all.

The result of such speculation is however slightly paradoxical. We seem to have arrived at an argument to the effect that we shouldn't believe our scientific theories to be true—yet the argument *itself* depends upon the results of some of those very same scientific theories! It is in fact our best theories of human development and cognitive psychology that seem to tell us that we should be skeptical of what our scientific theories have to say, evolutionary biology and cognitive psychology included. We seem to have found ourselves in a bit of a muddle.

So how do we reconcile these states of affairs? The philosopher Alvin Plantinga has made an intriguing suggestion in this respect.[9] He argues that if we want to take our scientific theories at face-value—including the belief that human beings evolved from simpler organisms with all the consequences that entails—then we need to have some guarantee that the cognitive heuristics leading to those scientific theories have not led us astray. And the only way we can be sure of that is if we suppose, like Sherlock Holmes, that someone has carefully planned things so that the base-rates are always in our favor. In short, Plantinga argues that religious belief is the only way to make sense of contemporary scientific success.

This is certainly a surprising conclusion. The common understanding is that science and religion must be in conflict, as evidenced by the clash between evolutionists and creationists. But Plantinga is not a creationist. On the contrary, he is a whole-hearted supporter of modern science. But he argues that if evolution is true, then we have reasons to believe that our cognitive faculties depend upon unreliable heuristics—and if our cognitive faculties rely upon unreliable heuristics, then it is extremely unlikely that any of the scientific theories we have produced are true. The only way out is to suppose that for all the bias and error in the human intellect, the world is somehow organized in such a way as for all these mistakes to balance each other out. And for Plantinga, that would quite literally require a miracle.

There may however be a more cynical conclusion to draw. Perhaps all that these various argumentative muddles show is that it is in fact very difficult to investigate the nature of science. Depending upon the level of abstraction, we have encountered arguments that rely upon our scientific methods to tell us that our scientific methods are reliable, and arguments that appeal to our scientific theories to tell us that we should not believe our scientific theories. In both cases, we have attempted to investigate the nature of science, while simultaneously acknowledging that it is science itself that provides us with the best guide to any form of investigation. The whole issue therefore runs around in a very narrow circle indeed, with the risk that we can only ever get out of these investigations what we are prepared to put into them at the beginning.

Science is our best way of investigating the world. If we were to come across a better method for doing so—reading tea leaves or gazing into a crystal ball—we would examine and investigate these methods, conduct experiments and double-blind trials, until they simply became another aspect of science. It is therefore hardly surprising that when we then attempt to uncover the deeper levels of these methods, we find ourselves repeatedly striking intellectual bedrock. The conclusion then might be that we cannot reasonably hope to provide an informative investigation of our scientific methods after all, since there simply is no other perspective from which such an investigation can be conducted. This is not a very exciting conclusion to be sure. But sometimes we make progress in the world, not by providing better answers, but by merely discovering which questions are worth asking.

5

Living in different worlds

In September 1905, Albert Einstein published his special theory of relativity. For a theory principally concerned with the velocity of light in a vacuum—roughly 700,000,000 mph as it happens—it was one with far reaching and rather surprising consequences. Perhaps most importantly, it entailed that our previously familiar notions of space and time were not in fact absolute and unchanging quantities as physicists from Aristotle to Newton had supposed, but needed to be significantly rethought. According to the special theory of relativity, the distance between two different objects, and the length of time something takes to happen, will actually vary from observer to observer, depending upon their own frame of reference. And this is not simply the fact that different people will often disagree about the facts, like the way colors can change under different lighting. According to the special theory of relativity for example, the faster you travel, *the slower the passage of time*, which means that one sure-fire way of fighting those wrinkles is to undertake an all-expenses paid intergalactic cruise at eye-watering velocities around the universe, since by the time you get back home you will have quite literally aged less than those of us who stayed at home.

Yet for all of its unintuitive consequences, the special theory of relativity was in many ways also a somewhat conservative proposal. It did not introduce any new data or experimental results, and nor did it postulate any new laws or mathematical principles. What Einstein offered instead was a novel way of thinking about old phenomena—an ingenious new framework for making sense of existing scientific results. More specifically, the special theory of relativity offered an innovative strategy for reconciling the long-established principles of mechanics and motion that had originated in the work of Galileo in the seventeenth century, with the more recent work in electromagnetism that had been established by James Clerk Maxwell at the end of the nineteenth century. In much the same way then that Newton had shown how both terrestrial and extraterrestrial phenomena could be incorporated within the same set of mechanical principles, so in turn did Einstein show how these familiar mechanical principles could themselves be combined into a single physical framework with the new-fangled notions of electricity and magnetism. While undoubtedly a work of genius, the special theory of relativity was also in many ways just a piece of scientific housekeeping, one that simplified our scientific understanding as much as expanding it.

The reaction to Einstein's theory was nevertheless surprisingly vigorous, and far from universally positive. There was in particular a great deal of resistance from within the scientific community itself, which just as in the case of Galileo 300 years before, demonstrated a remarkable resistance to anything that challenged their own cherished world view (or indeed, their reputation and continued funding). Many scientists thus complained that Einstein's proposal was unnecessarily radical, and at odds with the basic principles of common sense. Some even complained that the theory involved far too much mathematics, and that it was therefore far too difficult for anyone to understand. The philosophical community was especially upset, having long since established on the basis of indubitable first principles and lengthy armchair introspection that space and time were indeed constant and unchanging as Newton had proposed. They were therefore somewhat taken aback by the suggestion—not to mention the subsequent empirical confirmation a few years later—that this was empirically false.

What was most amazing, however, was the extent to which the special theory of relativity managed to elicit both enthusiastic discussion, and vitriolic denunciation, from the general public. While exposure to his original paper on the special theory of relativity had remained largely confined to other scientists and academic specialists, by 1915 Einstein had published his general theory of relativity—extending the results of the special theory to cover both accelerated motion and the principles of gravitational attraction—and the ideas slowly began to spread through the wider public. Writing to his friend and collaborator Marcel Grossman in 1920, Einstein observed that:

> The world is a strange madhouse. Currently, every coachman and every waiter is debating whether relativity theory is correct. Belief in this matter depends on political party affiliation.[1]

In retrospect perhaps, this situation was not so difficult to understand. The postwar period in Germany was naturally enough one of enormous social unrest. For the politically progressive, the theory of relativity came as a welcome breath of fresh air. It overturned existing conventions and opened up exciting possibilities for the future. It was quite literally a brave new world, a break from the past. For the more conservative, however, still smarting from Germany's defeat in the recent war and desperately looking for someone to blame, it represented everything that was wrong with contemporary society. They detested the moral decadence and artistic experimentation so characteristic of the Weimar Republic, and saw the proposed relativity of space and time as just another aspect of its wanton abandonment of reason, order, and (one assumes) traditional family values. To make matters worse, Einstein himself was an outspoken pacifist and social democrat—precisely the sort of person committed to undermining the national interest—and of course part of the ever present and predictably absurd Worldwide Jewish Conspiracy which had caused all of these problems in the first place.

All of these concerns found their most vehement expression in the curious figure of Paul Weyland, an individual now all but lost to the history of science. An engineer from Berlin, he claimed to possess a doctorate in chemistry, although there is no evidence that he ever attended university, nor for that matter, even graduated from high school. He was, however, undoubtedly the President of the racially unambiguous Association of German Natural Scientists for the Preservation of Pure Science—although then again, he also appeared to have been its only member. In 1920, Weyland achieved brief notoriety when he successfully packed the Berlin Concert Hall with a conference explicitly devoted to the denunciation of Einstein's theory of relativity. Delivering the keynote address, Weyland argued alternately that the theory was false; possibly true although limited in its understanding; almost certainly false; largely true but clearly plagiarized from his own work; definitely false; and ultimately too incoherent for anyone to be able to tell one way or the other (all of which is actually quite a lot like academic peer-review today). In any case, Weyland concluded, the theory of relativity had only ever come to prominence in the first place because Einstein's cronies in the Worldwide Jewish Conspiracy also controlled the mainstream media, and had embarked upon some nefarious scheme to mislead good honest citizens with their vile propaganda.

This conference was to mark the highpoint of Paul Weyland's scientific career. Throughout the rest of the 1920s, he edited an anti-Semitic journal—the imaginatively entitled *German Folk Monthly*—and published a moderately successful historical potboiler recounting the righteous slaughter of bloodthirsty Slavs by heroic German Knights in the tenth century. A second book outlining the moral dangers of dancing was advertised, but sadly never made it to print. Ever the entrepreneur, Weyland traveled to New York to sell his own special recipe for distilling motor oil from raw materials, and when that failed, to Stockholm to sell the same product as insecticide. He attempted to swindle the Norwegian Government into funding a spurious scientific expedition to the Arctic, spent several years on holiday in South America pretending to research tropical diseases, and provoked a minor international incident when he tried to invoke diplomatic immunity rather than pay his hotel bill in Zurich. All in all, Weyland was convicted three times for fraud, and was considered to be such a liability that despite his impeccable anti-Semitic credentials, the Nazi Party refused his membership application, and eventually even revoked his citizenship. Demonstrating an extraordinary degree of chutzpah—and a quite spectacular hypocrisy—Weyland immediately moved to Spain claiming to be the victim of political persecution, and survived for several years on the charity of precisely the kind of Jewish refugee he himself had helped to drive out of Germany. In 1938, Weyland moved to Austria. Ironically enough, a few months later the *Anschluss* was declared unifying Nazi Germany and Austria, and Weyland watched in horror as the Wehrmacht marched into the streets of Vienna. Still officially classified as a political criminal, Weyland was promptly arrested, and spent the entirety of the Second World War in a concentration camp at Dachau.

But you can't keep a good man down—or a bad one, for that matter—and there was no way a little hiccup like this was going to stop an embittered crackpot like Paul Weyland. Liberated by the Allies in 1945, he was able to work as an interpreter for the U.S. Forces, and gradually wormed his way into favor with the counter-intelligence community. Never one to overlook a golden opportunity when he found one, Weyland used his position to intimidate and blackmail innocent citizens, threatening to denounce them as Nazi sympathisers unless they paid up. In 1948, Weyland emigrated to the United States, where he was finally able to resume his lifelong passion and promptly denounced fellow emigre Albert Einstein as a communist to the FBI. The resulting investigation—while ultimately fruitless—produced nearly 1,500 pages of notes and speculations, to which Weyland was a proud contributor. In 1967, Weyland returned to Germany to take advantage of its socialist health care, and in 1972 died of heart disease at the age of eight-four. It was, I suppose, a full life.[2]

Yet for all of Weyland's faults, he might have had a point. I don't mean to suggest that Einstein was a fraud or a plagiarist or that some Worldwide Jewish Conspiracy controls the media. But the idea that the theory of relativity owed its success to social and political reasons more so than traditional scientific criteria is not without some degree of credibility. After all, the theory was not based upon any novel phenomena or observable effects, but simply offered a new framework for understanding existing scientific data. Later of course some degree of experimental confirmation did become possible, such as Arthur Eddington's expedition in 1919 to observe the bending of light during an eclipse as predicted by the general theory of relativity, and which did so much to bring Einstein's work into the public consciousness (and which so impressed a young Karl Popper growing up in Vienna). And today of course we can actually demonstrate the effects of time dilation through more precise means, such as by comparing the rates of decay of unstable subatomic particles at rest with those racing around a particle accelerator at close to the speed of light—the faster the particles travel, the slower the passage of time, and the longer they take to decay—or by calibrating a pair of unimaginably precise atomic clocks and sending one of them around the world a few times on a supersonic jet. But at the beginning of the twentieth century, few of these techniques were available or well understood, and yet everyone had an opinion. As Einstein himself noted, acceptance of the theory appeared to be determined as much by "party affiliation" as any considerations relating to improved predictive power or novel experimental result. This is also a view that has attracted considerable enthusiasm in many modern university departments, particularly amongst the more postmodern academics keen to emphasize the political undercurrents of even the most mundane aspects of our lives. And if that really is the case, if acceptance of a scientific theory has more to do with political conviction than interrogation of the evidence, then that raises some very significant questions about the scientific method.

In search of the aether

It might be helpful before we proceed to look a little closer at some of the key ideas behind Einstein's theory of relativity, and in particular, just why it managed to elicit the sort of conceptual horror that it did amongst individuals like Paul Weyland. In the nineteenth century, the Scottish physicist James Clerk Maxwell demonstrated that light was a special type of wave—part of the electromagnetic spectrum that runs from x-rays and microwaves at one end, through the various colors of the visible spectrum, and up to ultraviolet radiation at the other—and went on to determine its velocity and other important properties. One issue that remained somewhat unresolved, however, was the medium through which waves of light were supposed to be transmitted. In the case of a sound wave, for instance, we know that it can be transmitted through a body of water or the air around us through (roughly speaking) the successive collision of individual molecules. But there does not appear to be any such physical medium in the case of light, which travels from distant stars through the reaches of outer space where any such molecules are in very short supply.

Maxwell proposed therefore that there must be another, as yet undiscovered medium facilitating the propagation of light. This was called the *luminiferous aether* and was supposed to permeate every single nook and cranny of the universe in order to accommodate the almost ubiquitous waves of light that surround us. It was precisely because the aether was so pervasive that explained why up to that moment no one had noticed its existence, in much the same way that one imagines that a fish is completely unaware of the water through which it swims. Nevertheless it stood to reason that if the aether did exist, then there must be some way of measuring it. In particular, since the Earth is constantly rotating on its axis in its orbit around the Sun, it would seem to follow that it must in fact be in constant motion with respect to this all-pervasive aether, which should in turn entail any number of experimental consequences. In 1888, two American physicists named Albert Michelson and Edward Morley set out to perform just such an experiment.

The underlying idea was relatively simple. While Maxwell had established that the velocity of a wave will be determined by the properties of the medium through which it is traveling—which is why for example sound travels faster in air than it does in a more densely packed medium like water—our judgment of the *relative* velocity of a wave will also depend upon our own motion through that medium. If we happen to be moving through the medium towards the source of the wave for instance, we will judge the wave to be moving with a greater velocity, whereas if we are moving through the medium away from the source of the wave we will judge it to be moving with a lesser velocity. In the case of a sound wave, this can be observed by the change in pitch of a police siren as the car hurtles towards us and then recedes into the distance. If light travels through some all-pervasive aether, it follows then that there should be some similarly observable variations to be detected, depending upon our own motion with respect to this hitherto mysterious substance. In the Michelson-Morley experiment, beams of light were refracted and sent at different angles to one another, where they would travel a short distance along the apparatus before being reflected back to their source. It followed that if the experimental apparatus really was moving with respect to the aether—as the constant orbit of the Earth would suggest—then the relative velocity of the various beams would have to be different, as they would have been traveling in different directions with respect to the aether. This would manifest itself as an interference pattern when the different beams of light were recombined, and by analyzing the extent of the interference and calculating backwards, it would be possible to determine the frame of reference at which the aether was at rest, and therefore by extension whether or not it really exists.

The experiment was a spectacular failure. No matter how accurately the results were analyzed, or how much the different beams of light were varied, absolutely no interference patterns could be detected. If light really was propagated through the all-pervasive luminiferous aether, it must have some very peculiar properties indeed, since it seemed to be impossible to ever tell if we were moving with respect to it. A number of ingenious proposals were subsequently suggested to try and explain away this anomalous result. One option was to suggest that the Earth somehow managed to "drag" the aether along with it as it orbited the Sun, like water caught in the wake of a ship, which was why we always seemed to be at rest with respect to it. It was however difficult to explain how this could in fact be the case, and indeed why waves of light coming from distant galaxies did not demonstrate the sort of disruption this strange state of affairs would cause. More imaginatively, it was also proposed that the aether could have an effect on the apparatus used to investigate it, and was able to systematically distort our measuring equipment as we moved through it. In the case of the Michelson-Morley experiment for example, not only would the different beams of light travel at different relative velocities as the Earth traveled through the aether, but so too would the experimental apparatus itself contract and expand as it traveled through the medium, such that the slower beam would also end up traveling a shorter distance, sufficient to cancel out any of the predicted interference. Such responses, however, only tended to raise more problems than they solved, not least because they only seemed to apply when scientists were unable to come up with a better explanation.

Einstein's proposal by contrast was to take the phenomena at face value. He rejected the idea of an all-pervasive aether altogether, or indeed any other medium through which light propagated. This was in many ways the simplest response to take to the Michelson-Morley experiments, and had the added advantage of abandoning any of the mysterious forces at work selectively deceiving scientists that it seemed proponents of the aether theory had to accept. But there were other, less intuitive consequences of the proposal. If there was no medium through which light propagated, then there was no sense in which the relative velocity of a beam of light could vary from observer to observer, since there was no sense in which different observers could be moving in different directions through that medium. The velocity of light was therefore a constant for every frame of reference. But this was to have further, even more unintuitive consequences. Much of our everyday experience is based upon the idea that the relative velocity of an object will vary depending upon our own rate of motion. If I run towards a speeding car for instance, while you run away from it, we will naturally enough come to different conclusions about how long it will take for the car to reach us. In this case the car will hit me first, whereas if you run fast enough in the opposite direction, it may never catch you up at all. It seems reasonable to suppose that the same must be true for light. If I run towards a beam of light, and you run away from it, then it seems that one and the same beam of light should illuminate me before it illuminates you. But this is precisely what Einstein's solution denies—even though I am running towards the light and you are running away, the very same beam of light will in fact approach us both at exactly the same velocity. And this is rather surprising. It is after all one thing to accept that you can never outrun a beam of light; it is however quite another to accept that, no matter how quickly you travel, the same beam of light will continue to catch you up at *exactly the same rate*. It is like the situation in one of those dreadful old horror movies, where no matter how fast the heroine flees, the shambling corpse of the reanimated serial killer always seems to be slowly gaining on her, or when bored millionaires hunt Jean-Claude Van Damme for sport through the graveyards of New Orleans.

And that is not the worst of it. When I run towards a speeding car and you run away from it, we will form very different judgments about its relative velocity, but we will nevertheless agree on how far the car has to travel before it hits one of us, and how long it will take for that to happen. But if the velocity of light is the same for all observers, then the only way in which everyone can agree on how fast one and the same beam of light is approaching them is if they *disagree* on the distance it needs to travel and the time taken to do so. On this account then, while the velocity of light remains the same for all observers, the familiar concepts of space and time suddenly become relative to one's point of view. For much of the early scientific community, that was just a stretch too far. Abandoning the luminiferous aether seemed therefore to entail abandoning much of the commonsense framework in which science had operated for centuries. And for individuals like Paul Weyland, it spelled complete conceptual—and not to mention moral—anarchy.

It is important, however, not to misunderstand these ideas. While it is certainly true that the theory of relativity thoroughly rejects any objective notion of space and time, it does not thereby abandon any objective coordinate system for describing physical events, and nor does it license the more general sense of anarchy that some have assumed. Rather, the theory of relativity simply ascends to a more abstract level of description, replacing the notions of space and time with the single notion of *spacetime*. And while the spatial and temporal distance between events may indeed vary from observer to observer, the single spacetime separation between them will not. In fact, what the theory of relativity shows us is that our familiar notions of space and time are just different ways of decomposing the more fundamental notion of spacetime. In much the same way that the physical distance between two objects can be described in different ways by different observers—up and to the left, down and to the right, depending upon one's point of view—without thereby disagreeing on the overall distance, so too can one and the same spacetime separation between two objects or events be described in different spatial and temporal combinations without thereby disagreeing over this more fundamental quantity. So while it is true to say that the theory of relativity abandons some of our familiar framework for describing the world, it also shows why these concepts are redundant, based as they are upon a superficial picture of the world.

Science from above, and science from below

According to Paul Weyland, the success of a scientific theory has very little to do with its predictive success or explanatory power, but is instead largely determined by broader social and political factors—the clandestine machinations of secretive cabals, the unconstitutional interference of shadowy government figures, and of course the Worldwide Jewish Conspiracy. This is because Paul Weyland was what professional philosophers refer to as a *complete idiot*. In the days before the information revolution, his conference in the Berlin Concert Hall was the equivalent of one of those disreputable internet chatrooms where you can go to discuss how Bigfoot shot Kennedy, and exchange grainy photographs of Neil Armstrong and Stanley Kubrick taking a break from their punishing filming schedule. No records survive of what Paul Weyland actually thought about the supposed moon landings of 1969, but chances are that he thought the Jews had something to do with that as well.

Conspiracy theories notwithstanding, there are however some much more plausible lines of argument for emphasizing the social and political factors governing our reasons for adopting one scientific theory over another. The first draws upon the various difficulties that we have already encountered in trying to specify some set of disinterested principles or purely logical rules that might constitute the scientific method. The idea that a good scientific theory is one that is falsifiable, for instance, fails adequately to distinguish between those cases of genuine scientific theories and those of pseudoscientific nonsense. Moreover, and perhaps more importantly, we have also seen that the extent to which a scientific theory can be said to be falsified by a piece of evidence often turns out to depend more upon our personal inclination to soldier on with a promising line of research than it does upon any precise logical considerations. The open-minded and observationally neutral assessment of our experimental results can be fundamentally shaped by our prior beliefs and social circumstances. Amassing ever greater stores of evidence does nothing to make a scientific conjecture more reliable, and in fact can sometimes seduce us into committing a whole range of basic statistical errors and simply reading our own preconceptions back into the data. Even our best explanations have relatively little to tell us about the approximate truth of our scientific theories, but rather further demonstrate our uninspiring cognitive origins. But if none of these traditional considerations offer a realistic understanding of the scientific method, then the prospect begins to suggest itself that there is no such thing as the scientific method after all. And if that is the case, then it does seem reasonable to suppose that the reasons we have for adopting one scientific theory over another must depend upon social and political factors, since there does not appear to be *anything else* that could be responsible.

Of course, just because we have so far failed to identify the essential elements of the scientific method, it does not necessarily follow that no such account is forthcoming. Maybe we just have to try harder. And once we do have such an account—a precise algorithm for generating reliable scientific conclusions out of empirical data—our choice of scientific theory will presumably be entirely determined with no possible room left for our social biases or ideological commitments. But even then, perhaps there is another argument lurking here, a more fundamental consideration in favor of acknowledging the intrinsic social and political dimensions of scientific practice. Suppose for the sake of argument that there is such a thing as the scientific method, a precise set of rules and principles of reasoning to which all successful scientific practice can in fact be reduced, and which in turn allows us sharply to distinguish between genuine science and pseudoscientific nonsense. The question to be asked, however, is how exactly our prescientific ancestors—previously sitting around in some idealized state of nature, spending their days happily worshipping trees and banging rocks together—could have first come together to coordinate their activity and thereby establish the sort of scientific community from which these principles originally emerged.

The answer might seem obvious, and indeed it seems easy enough to imagine the sort of scenario where a number of scientifically like-minded individuals, naturally recognizing that they were all engaged in a similar sort of enterprise, first get together and begin pooling their resources and expertise in the pursuit of a shared goal. But this only raises another question in turn, namely how these protoscientists are able to recognize one another as being engaged in a similar sort of enterprise without *already* having some grasp of the scientific method. The problem is that, on the whole, scientists do not have all that much in common. There is an enormous difference in approach and outlook across the sciences, from the theoretical branches of the physical sciences to the more hands-on end of the biological sciences, and without even getting into the more murky waters of the social sciences. Even at a more practical level, there is a great deal of difference between all of the various groups of individuals we refer to as scientists. They do not all work in labs and relatively few of them wear white coats. Some conduct complicated-looking experiments with bubbling test tubes and sparking electrical equipment, while others just tend to endlessly scrawl equations across white boards. Some scientists are driven by the pure pursuit of truth whereas others—like any other human activity—are just simply trying to pay the rent. Indeed, the only thing that really seems to bind all of these disparate activities together is the belief that they all somehow exemplify a unique and distinctive methodology, a precise set of rules and principles of reasoning for investigating the world around us. But now we seem to have just argued ourselves into a circle, for if it is only on the basis of a shared set of methodological principles that our protoscientific ancestors could recognize one another as being engaged in the same sort of activity, then it seems that there must have been a working understanding of scientific practice *before* all of these individuals coalesced to establish the nascent scientific community. We seem to be left in the paradoxical position where we would have had to already know what science was before we could have invented it.

The conclusion seems to be then that whatever it was that first brought the nascent scientific community together, it must have been something *other than* a shared set of scientific principles—which leaves us again with some broader social or political motivations. The point then is that however it is that scientific activity is fundamentally organized, it is not through some kind of top-down set of directives delivered from above. We may well be able to formulate a set of rules and argue that this alone constitutes the scientific method—but unless our audience is already committed to the practice of science, there is no particular reason why they should care one way or another, let alone agree to adopt them. It follows then that any discussion concerning how we should formulate the scientific method must actually *presuppose* the existence of some kind of scientific community within which such a discussion can take place. And that means that there must be some understanding of science, some basic organizing principle, that precedes any of our abstract, intellectual discussion about falsification, induction, explanation, and all of the rest. The argument then is that however good our account of the scientific method, and however compelling our reasons for adopting one scientific theory over another on the basis of that method, this whole intellectual structure fundamentally depends upon whatever social and political mechanism originally brought people together in such a way that they could meaningfully think of what they were doing as science.

The foregoing argument is admittedly a little on the conceptual side, but it is part of an important family of arguments that have shaped much of the intellectual development of the twentieth century. One particularly good example concerns the development of spoken language. It might be initially tempting to suppose that a language evolves through some kind of collective decision, individuals sitting around and deciding that the four-legged animal that barks is to be called a "dog," and that the one with whiskers is a "cat," and so on and so forth. A moment's reflection, however, convinces us that such a scenario is completely and utterly hopeless, since in order for these prelinguistic individuals to come to such an agreement, they must already be able to communicate with one another, which is of course precisely the problem that our hypothetical naming ceremony was supposed to solve.[3] More generally, as the philosopher Ludwig Wittgenstein was at pains to point out, it is always difficult to understand how any form of social coordination could arise through a process of laying down rules, since unless we already enjoy some form of social coordination, we would never be able to agree on how those rules were supposed to be understood.[4] The moral of the story again then is that, while it may indeed be possible to identify rules and principles governing the use of language or the practice of science, we argue ourselves into a circle if we suppose that these rules and principles could have been used to *establish* the coordinated activity that they describe. For that, we need to look deeper into the social and political factors that brought our prelinguistic or prescientific ancestors together in the first place.

Indeed, there is something almost ironic in the lingering conviction that we can explain our scientific practice in terms of some set of rules or principles of method imposed from above, rather than in terms of some kind of sociopolitical coordination bubbling up from below. We no longer attempt to explain biological complexity in terms of some grand design laid down for us by God—or at least, *most of us* no longer attempt to explain biological complexity in terms of some grand design laid down for us by God—but rather as the largely accidental result of many smaller instances of genetic variation and environmental selection. Yet we often seem unwilling to apply what we consider to be a paradigmatic example of a good scientific explanation to the practice of science itself. However exactly science works, it did not begin by an explicit intention to realize this or that set of methodological rules or inferential principles. It began through individuals working together in a variety of different ways, some of which managed to survive better than others. As the practice continues, it becomes possible to reflect upon it and—perhaps—to distill certain general characteristics that we might suppose constitute the scientific method. But any such analysis will always rest upon the underlying social and political factors that made such coordinated activity possible, and which will continue to influence and shape the continuing practice of science.

Paradigms, progress, and other problems

The question as to how exactly scientific activity could be coordinated in the absence of any overarching set of rules or principles is the central occupation of Thomas Kuhn's *The Structure of Scientific Revolutions*, arguably one of the most influential books of the twentieth century and a work of philosophy that rivals even Karl Popper's much loved *The Logic of Scientific Discovery* in terms of its significance for the public understanding of science. It therefore goes without saying that you would be hard-pressed to find any (worthwhile) undergraduate course in the philosophy of science today that did not prescribe both of these texts on its reading list. However, while the academic reception of Popper's work is fairly uniform—good on the broad strokes, hopeless on the detail—I think it is fair to say that Kuhn's contribution has provoked a considerable range of interpretations and evaluations, and still remains a matter of some controversy.

At the heart of Kuhn's account is the idea of a *paradigm*. Roughly speaking, this can be thought of as a shared set of assumptions about how to investigate the domain under question—what sorts of questions to ask, what sorts of techniques to use, and what sorts of observations and evidence are to be considered relevant. It is important to note, however, that these various assumptions all add up to something far less than a fully-fledged scientific theory, and may even include a great deal of fundamental disagreement about what sorts of entities exist, or what sorts of laws govern their behavior. What they do provide, however, is a broadly accepted framework in which these disagreements can take place without the entire discipline fragmenting into isolated sects. Kuhn writes:

> Aristotle's *Physica*, Ptolemy's *Almagest*, Newton's *Principia* and *Opticks*, Franklin's *Electricity*, Lavoisier's *Chemistry*, and Lyell's *Geology*—these and many other works served for a time implicitly to define the legitimate problems and methods of a research field for succeeding generations of practitioners. They were able to do so because they shared two essential characteristics. Their achievement was sufficiently unprecedented to attract an enduring group of adherents away from competing modes of scientific activity. Simultaneously, it was sufficiently open-ended to leave all sorts of problems for the redefined group of practitioners to resolve.[5]

Paradigms function, therefore, because other potential scientists naturally recognize them as an example of what they too wish to achieve and emulate in their own scientific practice, although there may well remain considerable disagreement as to *how* exactly that should take place.

Usually, a paradigm will itself be formed around a particularly noteworthy result (what Kuhn would later come to call an *exemplar*). This might be a new and unexpected discovery, a considerably more elegant or ingenious method of producing a known phenomenon under laboratory conditions, or even just a more pleasing mathematical framework for treating an old problem. This last one is illustrated in the case of Copernicus, who as we have already seen failed to predict any new astronomical observations, or indeed managed to provide any meaningful degree of simplicity over the rival Ptolemaic model. What Copernicus did do, however, was to apply a host of novel mathematical techniques that caught the eye of younger scientists like Galileo. What paradigms therefore do is provide a way of coordinating scientific activity in the absence of any explicit rules. And this provides the answer to our question. We have already seen that it is impossible to define scientific activity into existence, since no matter how straightforward and precise our instructions, there is no way to guarantee that everybody would interpret these rules in the same way unless they were *already* operating as a largely coordinated scientific community. Any such attempt to build science from first principles actually presupposes what it is attempting to achieve. By contrast, a paradigm is something that disparate investigators can come to agree upon in their prescientific state of nature. It is a result sufficiently surprising, powerful, or elegant that many different individuals can agree upon its importance without already sharing an underlying methodological commitment. Of course, there will be disagreement as to just *why* the paradigm is important, and it may well require extended discussion and hard work before any serious coordinated activity comes about. The important point is that we have at least a starting point for these various disagreements. And once a paradigm is accepted, practicing scientists no longer have to begin every investigation from scratch, but can begin to address themselves to others who share enough of their fundamental assumptions to not require lengthy introductions to the subject. Research becomes more focused, journal articles become more difficult for the layman to follow, and progress (usually) begins apace.

There is, however, an important consequence of this understanding of scientific practice. Since much of scientific activity will be shaped and guided by the shared paradigm, much of the actual work performed will be devoted simply to articulating that paradigm—confirming already well-known results, replicating existing experiments with ever so slightly improved degrees of accuracy, finding mathematically more elegant ways of expressing the known facts. This is what Kuhn calls *normal science*, the day-to-day activity that is the experience of the vast majority of professional scientists going about their nine-to-five existence. It is in many ways the polar opposite of what Popper would have characterized as scientific activity, neither risky nor imaginative, and certainly not conducted in any expectation of falsification. It is the sort of slow and steady progress that results from countless little cogs slowly rotating in the larger machine.

The problem lies in when things start to go wrong. For the vast majority of scientific practitioners, working within a well-articulated paradigm that simultaneously defines what they understand to be scientific practice, there is no room for error. If the experiment does not produce the expected outcome, if the mathematics do not balance out at the end, there remains little recourse but to blame the graduate student and do it again, or to shelve the anomaly for another occasion. If a theory really was to be falsified in the Popperian sense, anarchy would result, for it is only because the scientific community recognizes a particular scientific result as worthy of emulation that it manages to exist as a scientific community. There is no mechanism—or at least, no *scientific* mechanism—for dealing with a falsified theory, since in the absence of that theory, there is no shared understanding of what it means to be scientific. No, much better to put the whole incident down to faulty equipment or too much coffee, and just get back to articulating the shared paradigm.

Of course, such an approach can only go on for so long. Eventually the anomalies start to pile up and become too numerous to ignore. Gradually faith begins to fail in the paradigm and scientific activity becomes less coordinated. Isolated individuals and small groups begin to think about new ways to articulate the existing paradigm, or propose a new paradigm altogether. The different approaches will be discussed and debated, and eventually one will come to overthrow the existing conventions, and a new scientific paradigm takes its place. Crucially, however, none of this discussion can be understood as what we might think of as a scientific discussion—since of course part of what is under discussion is just what it means to be scientific in the first place. As Kuhn puts it:

> Like the choice between competing political institutions, that between competing paradigms proves to be a choice between incompatible modes of community life. Because it has that character, the choice is not and cannot be determined merely by the evaluative procedures characteristic of normal science, for these depend in part upon a particular paradigm, and that paradigm is at issue. When paradigms enter, as they must, into

a debate about paradigm choice, their role is necessarily circular. Each group uses its own paradigm to argue in that paradigm's defense.[6]

If a paradigm defines what constitutes scientific activity, then any debate over the adoption of a paradigm must by definition be nonscientific. Such arguments will be based on aesthetic values orthogonal to the experimental data, or perhaps the personalities and reputations of the scientists in question. Or perhaps other methods of persuasion will come to the fore—political ideology, the promise of larger funding grants, threats of hiring and firing.

But now we seem to be very close to where we began. Paul Weyland denounced the success of Einstein's theory of relativity as nothing more than propaganda and the sinister machinations of the media. We dismissed these conspiracy theories as the ravings of an anti-Semitic loon, but did acknowledge that any plausible account of scientific practice would have to acknowledge some role played by broader social and political considerations. In particular, we noted that scientific activity presupposes the existence of a shared scientific community in which largely shared values and attitudes could be articulated, and noted that whatever it is that holds a scientific community together cannot itself be scientific. This idea is further articulated by Thomas Kuhn in terms of a paradigm that holds the community together. But in following through the idea of a paradigm, we seem to acknowledge the fact that any choice of paradigm must itself be determined by broader social and political factors, propaganda, and mob psychology. Perhaps Weyland was right after all.

Relativism and its discontents

At the heart of Paul Weyland's dissatisfaction with Einstein's work—besides the rampant anti-Semitism of course—was the notion that the relativity of space and time that followed from rejecting the luminiferous aether somehow entailed a more thoroughgoing relativity of all other social values. It was the idea that once we allow such familiar notions as the distance between two objects or the time taken for an event to happen to vary from observer to observer, then it did not seem a far step to the idea that other familiar notions such as right and wrong, or the difference between true and false, might similarly vary upon your particular point of view. The inference is of course completely bogus. On the one hand of course, there is no straightforward connection between a scientific view about space and time and one's particular moral perspective. And on the other hand, the special and general theories of relativity do not entail that all spatial-temporal frameworks are up for grabs, but rather replaces the old notions of space and time for the more general notion of spacetime.

Nevertheless, by following through some of the more credible elements of Weyland's concerns, we have found ourselves again facing a more wholesale relativism. We have seen a number of reasons for supposing that the acceptance of a scientific theory may well depend on broader social and political factors—whether out of despair at finding any other factors upon which to base our choice, or through acknowledging that there must be some nonscientific factors behind the development of a scientific world view in the first place. This line of thought has seen its most well-known development in the work of Thomas Kuhn, who has sketched some of the ways in which everyday science can be shaped around largely unarticulated exemplars and paradigms, rather than explicit sets of rules or principles. But it seems to be a consequence of Kuhn's view that, if the standards of scientific evaluation are determined by one's current paradigm, then any change of paradigm must necessarily be based on solidly nonscientific factors. We seem to be in the position again where the very standards we have for cool and rational appraisal may well owe themselves to baser, sociopolitical aspirations. Notions of scientific right and wrong will depend upon which society or culture you happen to find yourself in.

This brand of conceptual relativism is certainly not new. The Ancient Greek Sophist Protagoras once famously declared that "man is the measure of all things" and traveled the country teaching that truth is relative and all beliefs are true. Such reasoning did not impress Socrates, who replied simply that he didn't believe him. When it was suggested that what Protagoras really meant was that all beliefs are true *from someone's point of view*, Socrates asked

FIGURE 5.1 *Einstein's theory of relativity left man adrift in a strange universe of an unimaginable scale, where the familiar notions of space and time broke down and the laws of nature seemed to run amok beyond our parochial corner of the galaxy. This sense of dislocation inspired a considerably less optimistic genre of science fiction, such as the work of H. P. Lovecraft illustrated above, where sentient fungi manipulated life on Earth, tentacled monstrosities waited to consume the stars, and grasping the true nature of existence could drive one to insanity. Although less hostile to the principles of relativity than Paul Weyland, Lovecraft did nevertheless share his rampant anti-Semitism, which seems to have been a recurring theme amongst the theory's commentators.*

whether or not this was a truth that everybody was supposed to accept? After all, Protagoras traveled the length and breadth of the country expounding this particular doctrine, so he obviously supposed it to have universal application. But if that was the case, then the very claim that everything is relative is precisely the sort of universal truth that relativists claim do not exist.[7]

In the more contemporary instance of incommensurable conceptual schemes, a similarly awkward question was posed by the American philosopher Donald Davidson.[8] He asked how it was that, if different paradigms or cultural backgrounds really do produce these mutually unintelligible perspectives on the world, how it was that their proponents were nevertheless able to *describe* them in such wonderful detail? He notes for example that Kuhn's earlier work on the Copernican revolution is specifically devoted to the task of trying to make clear the almost incomprehensible strangeness of the pre-Copernican paradigm—a world of Aristotelian objects moving under their own natural inclinations, as part of a grand cosmic hierarchy where man occupies the unmoving center of both a physically and morally ordered universe, and where the prospect of a moving Earth stood in stark opposition to both our everyday experience and our theological understanding of ourselves. The problem, however, is that this is a world view that Kuhn not only conveys in outstanding detail and depth throughout the course of his book, but one that he evidently has no difficulties expressing entirely with a modern, post-Copernican idiom!

This line of thought can be developed into a general challenge for any prospective relativist. In order to take the theory seriously, we would naturally enough wish to be given a concrete example of how different people and different cultures possess incommensurable conceptual schemes. If no such example is forthcoming, then the claim can be dismissed as so much hyperbole. And if such an example is presented, then the very fact that these supposedly incommensurable conceptual schemes have been sufficiently well compared so as to establish their incommensurability only shows that ... well, that they weren't really incommensurable in the first place.

The idea that different scientists working with different scientific paradigms are therefore working with radically different conceptual

LIVING IN DIFFERENT WORLDS

schemes is at best impossible to establish—and at worst, straightforwardly incoherent. That does not mean, however, that Kuhn's central contention is wrong. We have already seen repeatedly throughout our discussion in this book that simple, monolithic conceptions of the scientific method are rarely straightforwardly applicable, and that different scientists working in different traditions will often make different judgments about the relative importance of different considerations. To take one simple example, we have seen that the purely logical process of subjecting our scientific theories to rigorous testing and falsification depends as much upon our willingness to make adjustments elsewhere in our theoretical world view as it does upon any objective relationship between theory and data. And the extent to which one scientist may be willing to make any particular adjustment to preserve a scientific theory from falsification will inevitable depend upon his other theoretical beliefs, but perhaps more importantly upon a less clearly articulated sense of when enough is enough. Questions as to how many additional planets can be posited, or whether other fundamental physical assumptions can be given up, in order to preserve Newtonian Mechanics from refutation will rarely be articulated in terms of a precise set of rules and criteria. It will depend upon an intuitive sense of when one is developing a promising scientific theory, and the point at which one is desperately salvaging a failed conjecture, which in turn will be based upon the scientist's paradigmatic examples of good scientific practice.

The point is, however, that one can acknowledge this more nuanced understanding of science without running screaming and shouting into full-blown conceptual relativism. Just because there is no single universal principle that characterizes every single instance of scientific practice, it does not follow that the entire process is random and arbitrary. It also does not mean that no comparisons are possible, or that one scientist working in one paradigm cannot criticize another scientist working in a different paradigm. For example, different scientists working in different paradigms might disagree as to which phenomena are worth explaining. After the Copernican revolution, which put the Earth in motion against a background of distant stars, there was an ongoing concern as to why the relative position of these distant stars did not appear to move as the Earth revolved around the Sun (the reason is that they are so distant that one requires an extraordinary powerful telescope to detect this parallax motion). For the Ptolemaic astronomer by contrast, who envisioned a stationary Earth at the center of a fixed heaven, there was no such motion to observe. Nevertheless, while Ptolemaic and Copernican astronomers might have disagreed as to what phenomenon needed to be explained, they could presumably agree on *how many* phenomena their respective theories could explain. Similarly, while different paradigms might impose different criteria concerning the accuracy and precision that different scientific theories should meet, there are nevertheless perfectly objective judgments to be made as to how well each paradigm manages to satisfy its own criteria.

But there is a more general point here. If two different people come to different conclusions about the same issue—and assuming of course that both people can advance good reasons for their conclusions—then we know that at least one of them must be wrong. If our own community encounters another society, with their own standards and paradigms of scientific practice, who endorse a radically different conception of the world, then we should be cautious. It would be pure chauvinism to suppose without further reflection that we are right and they are wrong. But a degree of humility in the face of conflicting opinions, and an open-minded attitude to other ways of seeing the world, is not the same thing as cultural relativism which holds—incoherently—that everyone is somehow equally right.

Moreover, there is an important practical difference between these two attitudes. If we suppose that our different views are in genuine conflict, and that at least one of us must be wrong, we are encouraged to reassess our beliefs and conduct further experiments in order to help determine which one of us it might be. On the humility model, conflicting paradigms is an impetus to progress. But for the conceptual relativist, everyone is right from their own point of view. A conflicting paradigm is therefore of no greater intellectual importance than the fact that some people prefer tea and some people prefer coffee. There is no impetus to further investigation, and no encouragement to seriously engage with the conflicting paradigm. While often championed as the more enlightened attitude, conceptual relativism in fact encourages a far more close-minded attitude towards our own beliefs and world views. It also encourages a more patronizing view towards other cultures, whose own potential contribution to the ongoing scientific process can be easily dismissed as "true for them" and thus without a claim to universal validity.

6

The bankruptcy of science

The future looks bleak. We are squandering our limited natural resources at an ever increasing rate. As the population grows, the demand will only accelerate, and soon there will be nothing left. Our mines will stand empty and our cities dark and silent, a testimony to our hubris and the extravagances of human folly. We must act now before it is too late, or else condemn future generations to untold suffering and hardship. We must not only restrain our rapacious appetites, but cut them down to a more manageable level. Industry must be rolled back so that we can live in harmony with nature. The very survival of the human species stands in the balance.

Or so wrote the English economist William Stanley Jevons in 1865. Unlike contemporary environmentalists, Jevons was alarmed not by rising surface temperatures, the disappearance of the rain forests or the gradual acidification of the oceans, but rather by the growing disparity that existed between the United Kingdom's burgeoning demand for coal and her ever dwindling reserves. In his imaginatively entitled *The Coal Question: An Inquiry Concerning the Progress of the Nation and the Probable Exhaustion of our Coal Mines*, Jevons predicted that on the basis of the current rate of growth, the United Kingdom would require over 100 billion tons of coal over the course of the twentieth century—an astronomically vast figure, and one that significantly outstripped even the most generous estimates of the U.K.'s total lifetime supply, or indeed that of the entire planet. It was in Jevon's view therefore nothing short of a disaster waiting to happen, and one that demanded in turn the most exceptional of measures in response, if not the complete renunciation of the entire

Industrial Revolution. In what amounted to an early anticipation of the principle of sustainability, Jevons wrote:

> We are growing rich and numerous upon a source of wealth of which the fertility does not yet apparently decrease with our demands upon it. Hence the uniform and extraordinary rate of growth which this country presents. We are like settlers spreading in a new country of which the boundaries are yet unknown and unfelt. But then I must point out the painful fact that such a rate of growth will before long render our consumption of coal comparable with the total supply. In the increasing depth and difficulty of coal mining we shall meet that vague, but inevitable boundary that will stop our progress ... There is, too, this most serious difference to be noted. A farm, however far pushed, will under proper cultivation continue to yield for ever a constant stop. But in a mine there is no reproduction and the product once pushed to the utmost will soon begin to fail and sink to zero. *So far, then, as our wealth and progress depend upon the superior command of coal, we must not only stop—we must go back.*[1]

But it was a matter that went far beyond the mere economic survival of the British Empire; for Jevons, it was an issue of "almost religious importance."

It should be noted, however, that the predicted depletion of our fuel supplies and the irrevocable collapse of society as we know it was not the only contribution that Jevons made to Victorian intellectual life. He was one of those great polymaths with an opinion on almost every field of human endeavor, and absolutely no compunction in their expression. An early pioneer in the field of microeconomics and the theory of marginal utility, Jevons was nevertheless repeatedly unlucky in the management of his own finances. Demonstrating an economist's cast-iron conviction in his own lack of culpability, Jevons set about investigating the possible causes of these otherwise inexplicable mishaps, and following a chance conversation with one of his colleagues in astronomy, came to the tenuous conclusion that the roughly ten-year cycle of his own bad investments must be due to the hitherto unappreciated influence of sun-spots on the global trade cycle. As a committed social reformer,

THE BANKRUPTCY OF SCIENCE

Jevons violently opposed the establishment of free hospitals and charitable health care, arguing that it undermined the character of the poor and encouraged a culture of dependency—but he did nevertheless advocate the wide-spread availability of concert halls and classical music, which presumably had a much more beneficial effect on the needy. Jevons even had strong views about the proper layout of museums, writing voluminously on the subject throughout his life. He deplored the eclectic mix of exhibits that still characterize many contemporary institutions, which he saw as not only superficial and without merit, but also as actively detrimental to one's intellectual development. Jevons argued that these cluttered displays encouraged the belief that one could become educated simply by wandering aimlessly throughout an exhibition with an open mouth and a vacant expression, and thereby undermining the habits of careful study and sustained concentration to which all should aspire. He recommended that school children be banned from museums altogether for the sake of their cognitive improvement.

Less than twenty years after the publication of *The Coal Question*, the United Kingdom's actual output of coal was already significantly less than the value Jevons had predicted. New sources of power had begun to take the place of coal, such as petroleum and electricity—a possibility which Jevons had considered in his book but dismissed outright as pure fantasy, something better suited to the fevered speculation of science fiction and pulp novels rather than serious academic consideration. The total production of coal between 1865–1965 was in fact just under 2 billion tons, less than 2 percent of what Jevons had originally predicted. By not taking into account the way in which the economy can change and develop as a result of technological development, Jevons was not just mistaken, he was *spectacularly* mistaken. Offering his own characteristically sardonic diagnosis of the situation, the influential economist John Maynard Keynes wrote:

> His conclusions were influenced, I suspect, by a psychological trait, unusually strong in him, which many other people share, a certain hoarding instinct, a readiness to be alarmed and excited by the idea of the exhaustion of resources ... Jevons held similar ideas as to the approaching scarcity of paper as a result of the

vastness of demand in relation to the supplies of suitable material (and here again he omitted to make adequate allowance for the progress of technical methods). Moreover, he acted on his fears and laid in such large stores not only of writing paper, but also of thin brown packing paper, that even today, more than fifty years after his death, his children have not used up the stock he left behind him of the latter; though his purchases seem to have been more in the nature of a speculation than for his personal use, since his own notes were mostly written on the backs of old envelopes and odd scraps of paper, of which the proper place was the waste-paper basket.[2]

A true individualist to the end, William Stanley Jevons drowned in 1882 after ignoring his doctor's orders not to swim.

What is perhaps most surprising, however, is that Jevons is merely part of a long and distinguished tradition of doomsayers whose predictions have turned out to be spectacularly wrong. In 1908, a National Conservation Commission established by U.S. President Theodore Roosevelt confidently predicted that supplies of natural gas would be completed exhausted by the 1930s, and oil reserves depleted by the 1950s. By the turn of the twenty-first century and the recent development of fracking technology, North America currently possesses the largest gas reserves on the planet. In 1968, American biologist Paul Ehrlich concluded that the "battle to feed humanity was lost" and that India was on the brink of total starvation. Today, there are over 1.2 billion Indians who would probably disagree. Never a man to avoid the most sensationalist of headlines, Ehrlich also predicted that all important ocean life would be dead by the 1980s, and that environmental hardship, rising sea-levels—not to mention the pervasive stink of dead fish washed up on the coast—meant that the United Kingdom would no longer exist as a political entity by the year 2000. I can only assume that most modern Britons have simply grown used to the smell. Writing in 1972, influential environmentalist Edward Goldsmith argued that industrialization was unsustainable, and that its "termination within the lifetime of someone today is inevitable." British economist and U.S. policy advisor Barbara Ward did not like the chances of humanity surviving beyond the year 2000. In 1990, climatologist Michael Oppenheimer predicted droughts and food riots across North America and Europe in 1995, and the Platte River in Nebraska drying up. Yet the most substantial food shortages of the twenty-first century have in fact occurred in Africa and the Middle East, and not from drought or other environmental catastrophes, but as an economic consequence of the vast areas of land in the West converted from food production to the development of biofuels—a bitterly ironic case of the purported solution causing the dangers it was supposed to avoid. In 2000, Oppenheimer also lamented that children may never play in snow again, which having recently spent a winter working on this book in upstate New York, remains one of the few sensationalist predictions that I am sorry to have seen thwarted.[3]

Given such a spectacularly poor track record of apocalyptic prediction, it is perhaps not so surprising that many commentators have expressed a pronounced skepticism about contemporary environmental warnings. They ask why we should take seriously the claim that global temperatures have been rising continually over the past century when only thirty years ago we were confidently anticipating the next ice age, or why we should be concerned about the imminent depletion of our natural resources when previous doomsayers have so consistently been shown to be wrong. Inevitably, of course, many of these challenges simply come down to matters of conservative politics, personal interest, or substantial financial investment in the fossil fuel industry—although conversely, we should not underestimate just how much money there is to be made in renewable energy subsidies and environmental protection grants, nor forget that nothing sells newspapers quite like the end of the world. But let us put these human foibles aside for the moment. More than enough ink has already been spilt on these particularly divisive tribal issues, and I have no intention of contributing further to what we might charitably refer to as the "debate" over the politicization of climate change. For somewhere amongst all of the eschatological anxiety and hyperbolic doubt lies a genuinely interesting philosophical question: whatever else one might think, given the mistakes of the past, is it not in fact *reasonable* to maintain a skeptical attitude about the present?

On the dangers of hindsight

What the preceding considerations would seem to suggest is that, since so many of our catastrophic environmental predictions have failed so spectacularly in the past, we should in fact be extremely skeptical about contemporary environmental predictions, be it manmade global warming, the melting of the polar ice-caps, the disappearance of the rain forest, or any other well-known variation upon the imminent apocalypse. It is an argument that appeals to our natural sense of caution—once bitten, twice shy—and to what is increasingly seen to be a healthy distrust of those in positions of authority. We no longer recognize the man in the white lab coat as some kind of disinterested oracle of truth, and we certainly do not trust the gaggle of failed politicians and supercilious celebrities who seem to make a very comfortable living out of peddling the latest doomsday scenario. It certainly does not help to be lectured on the evils of air travel by individuals flown in specially for the occasion, or told that we must preserve our painfully limited resources by someone who maintains a fully-staffed superyacht larger than the average family home for the occasional weekend jolly. But perhaps more importantly, it is also intellectually rather satisfying, not least because it appears to constitute what we might think of as a *scientific argument* against the latest pessimistic prophesy. The bulk of this book has of course been devoted to undermining the idea that there is such a thing as a unique and specific scientific method, but there are certainly rough and ready rules-of-thumb, and one of the things that science tells us is that the more often something has happened in the past, the more likely it is to happen in the future. As we have seen, the process is not flawless, and nor is it without important exceptions, but the idea that we should begin with a number of specific instances and extrapolate to find a general pattern must nevertheless remain an element of the scientific process. Thus when it comes to predictions concerning our ever-changing climate or rapidly depleting resources, the fact that we have been so badly wrong in the past does suggest that it is only rational to suppose that we will continue to be wrong in the future.

There are, of course, a number of different responses that could be made at this point. We might argue, for example, that it is simply unfair to compare the predictions of contemporary scientists to the speculations of historical individuals like William Stanley Jevons, who for all his ingenuity was clearly working in a period with a significantly impoverished scientific understanding. He did not have access to anything like the range of data available to the modern scientist, nor the kinds of sophisticated computer models for analyzing it. Neither for that matter was he a member of the sort of international research networks that characterize much contemporary investigation, with the corresponding degree of collaboration, peer review, and division of intellectual labor. On a more practical level, Jevons would hardly have had any graduate students to run errands for him, everything would have been handwritten, and in the United Kingdom at least it would take another century before you could get anything approaching a decent cup of coffee. It is therefore not surprising that previous predictions have proven to be so unsuccessful given the circumstances, but this can in no way impugn our most up-to-date analyses.

Such a response has some merit, but it only raises another question in turn. It is one thing to suppose that we have good reasons to believe that contemporary scientists are more reliable than their predecessors, and that therefore we have good reasons to trust their predictions—but how do we know that future generations will not make the same unfavorable assessment of us? At the time he was writing, William Stanley Jevons was one of the leading economists of the day, one of the founders of the field of microeconomics and an important influence on other important economists like Alfred Marshall and (despite all his withering sarcasm) John Maynard Keynes. For contemporary Victorians, there were good reasons to suppose that he was more reliable than his predecessors, and that therefore they had good reasons to trust his predictions. They would, in fact, have been able to advance *precisely the same arguments* for trusting their contemporary predictions about the imminent depletion of our natural resources as we are today. And they were wrong. How then can we be so sure that our position here at the beginning of the twenty-first century is such an unprecedented golden age of scientific reliability that future historians will never look back on us with the same wry humor with which we treat William Stanley Jevons?

There is therefore some initial plausibility to the idea that we should reject our contemporary environmental predictions on the grounds that they have been so misguided in the past. But, paradoxically enough, the real weakness of the argument might just be that it is *too strong*. One of the most intriguing features of the foregoing reasoning is how readily it generalizes to other aspects of scientific practice. After all, the fields of environmental science are not the only areas of scientific enquiry that have suffered severe setbacks or been the subject of significant revision in the face of inaccurate prediction. In medieval astronomy, it was believed that the planets rested upon vast crystalline spheres rotating slowly in the heavens, set in motion by a divine nudge of the elbow in accordance with some ineffable mathematical design. By the end of the seventeenth century, Newton has set the planets free to orbit the Sun in a graceful ellipse through the empty expanse of space, held in place by an unmediated and mysterious force that acted like some kind of cosmic centrifuge. Later, with the confirmation of the general theory of relativity, the planets found themselves constrained by neither crystalline sphere nor gravitational force, but merely by the intrinsic shape of spacetime itself, sagging into a great astronomical trough around the enormous mass of the Sun. And any confidence we have regarding the approximate truth of our current understanding of the universe can in turn only add to our conviction that previous astronomical theories were univocally false.

Moreover, the list is easily extended. In the realm of optics, light was originally supposed to be composed of tiny particles or corpuscles, traveling through the air at different speeds according to their intensity, until finally colliding with the human eye and impacting upon the optic nerves. This was superseded by the supposition that light was a wave, a series of vibrations in an all-pervasive luminiferous aether, a sort of invisible fluid filling the universe and allowing light to propagate like waves in an ocean. James Clerk Maxwell argued that light was a type of electromagnetic radiation, and Einstein maintained that the aether was redundant. According to the modern quantum theory, light sometimes behaves like a particle, and sometimes behaves like a wave, and confusingly enough, sometimes behaves like both at the same time. Chemical combustion used to release phlogiston into the atmosphere, a gas that was held to possess a negative weight in order to explain why some metals get heavier when they are heated; hot bodies became cold by losing their caloric; tiny organisms could spontaneously generate from the atmosphere alone; the human body was animated by irreducible vital forces or animal spirits, directed through the pineal gland ...

The fact of the matter is that once we start casting a critical eye over the history of science, it can begin to look like nothing more than an endless succession of just one failure after another. Almost every scientific theory that we have ever held has eventually been abandoned—and we are not just talking about superstitious crackpots back in the mists of time, or short-lived conjectures that no one ever really took seriously. Great thinkers like Copernicus and Galileo were eventually shown to be wrong. Isaac Newton's theories of motion were so successful as to go unchallenged for nearly 300 years, leading contemporary physicists to complain that there was nothing left for them to discover, before finally being consigned to the dustbin of history. It is not an encouraging spectacle. Henri Poincaré, the great French mathematician and physicist who helped to establish the mathematical foundations for Einstein's theory of relativity, summed up the situation with his own characteristic flair:

> The man of the world is struck to see how ephemeral scientific theories are. After some years of prosperity, he sees them successively abandoned; he sees ruins accumulated upon ruins; he predicts that the theories in vogue today will in a short time succumb in their turn, and he concludes that they are absolutely in vain. This is what he calls the bankruptcy of science.[4]

Climate change then is only the tip of the supposed rapidly melting iceberg. Once we start considering the history of science, it looks as if we should doubt *every single scientific theory* that we currently accept.

But that is surely a step too far. For even if we suppose for the sake of argument that the scientific study of climate has been nothing more than a litany of failure—and there are, of course, plenty of reasons why we might reject that starting point—that does not seem like a good reason to thereby reject great swathes

FIGURE 6.1 *A famous philosopher of science once summed up his entire position to me with the above illustration of the Temptation of St. Anthony. The point was that we can happily accept that St. Anthony was a real individual, and that he suffered various torments in the desert, while treating the fabulous beasts and demons lifting him into the air as largely illustrative. In much the same way, we can happily endorse parts of our scientific theories—the claims about observable phenomena, the underlying mathematical equations—while similarly withholding belief in some of the more esoteric theoretical detail in which they are presented.*

of physics, chemistry and engineering. Yet that is precisely what the foregoing line of argument seems to recommend. For if our theories of climate change and environmental exhaustion have suffered a few bumps along the road, then so too has almost every other scientific conjecture in the history of mankind. The argument from history simply proves far too much, which is usually a very good indication that there is a serious logical flaw somewhere in its reasoning. Philosophical reflection is rarely that effective. So what exactly has gone wrong?

The exquisite art of epistemological judo

To put it in simple terms, the argument from history is an attempt to use the strength of science against itself. There are many ways in which this might be done. It is certainly the case that since its supposedly humble beginnings in the work of individual geniuses like Galileo and Newton, science has grown and expanded into a vast and extremely well-funded enterprise, with increasingly close relationships to industry, the government, and the military. It has in short become part of "the establishment"—no longer engaged in the disinterested pursuit of truth, but pursuing its own shadowy agenda. This is the sort of argument that enjoyed some popularity in the 1960s in the opposition to what was seen as the complicity of the scientific establishment in nuclear proliferation during the Cold War, and more recently in response to a perceived bias in favor of an unwarranted—but financially lucrative—apocalyptic environmental world view. On this view, the vast funding resources controlled by scientists are seen not so much as a consequence of their unparalleled success, but as proof of an ulterior motive; the sprawling interconnected research networks indicating not so much the global scope and reach of scientific enquiry, but of conspiracy. In other words, it is the very success of science and all of the resultant trappings of its enhanced status that gives us reasons to be suspicious of its results.

Fortunately, this is not the sort of argument I have in mind. While the suspicion of some sort of hidden agenda remains an element of much scientific skepticism, there is a much more interesting way in which to use the strength of science against itself. It is possible instead to appeal to the actual methods and strategies of scientific practice to produce an argument against science. We have already encountered this sort of strategy in our discussion of creationism and the legal issues surrounding its introduction into the high-school curriculum with which this book began. Having initially occupied a strong position in the 1920s—when the teaching of evolution was something that could be prosecuted in the courts—creationists found themselves fighting an increasingly defensive battle throughout the rest of the twentieth century. By the 1980s the strategy was to demand the "balanced treatment" of creationism and evolution, on the grounds that it was *good scientific practice* to consider all of the relevant alternatives. It was therefore science itself that demanded that creationism be given a fair hearing.

In the case of creationism, however, such a strategy was of course simply disingenuous. The goal was not to secure the balanced treatment of relevant scientific alternatives, but rather to promote the explicit endorsement of a specific religious doctrine— and it is this that demonstrates the fundamental incoherence of the entire strategy. Ultimately, the idea is to appeal to the open-minded spirit of scientific investigation in order to justify a view about the origins of life that *does not itself* endorse the impartial consideration of competing points of view. It is therefore an attempt to use the central principles of scientific practice to support a position that simultaneously rejects those very same principles. But you can't have it both ways. If the underlying motivation for the "balanced treatment" of creationism and evolution is the open-minded spirit of scientific investigation, one cannot then endorse the view that biological complexity arose through a process of spontaneous creation without extending the same courtesy to every other possible explanation for the same state of affairs. Flipping the argument on its head, we could just as easily argue that if the creationist is going to reject the idea that alternative explanations for the origins of life are to be given a "balanced treatment," then there are no compelling reasons why the scientific establishment should give him a fair hearing in turn.[5]

Such an argument is a good example of what my Ph.D. supervisor used to call *epistemological judo*. The idea is that, whereas the martial artist would usually try to use the greater physical strength and momentum of an opponent as a way to unbalance and defeat them, so too does the creationist try to use the greater *intellectual virtues* inherent in good scientific practice as a way to undermine it—not so much the honest and in your face Jean-Claude Van Damme spinning back-kick of an argument, but more the Steven Seagal-inspired wave your hands and let your opponent fall out through a window style of argument. Thus it is because science demands the open-minded assessment of the relevant alternatives that we are supposedly compelled to give equal consideration to an explanation for the origins of life, even one that otherwise abjures all other principles of our scientific methodology. And the same holds for the argument from history. We all agree that it is good scientific practice to extrapolate from past instances in order to make predictions about the future. Therefore—according to the argument from history—it must be good scientific practice to extrapolate from the past failure of our environmental predictions to the future unreliability of contemporary worries about our diminishing natural resources and rising global temperatures.

But this is clearly another technique taken from the epistemological judo handbook, another attempt to use our scientific methodology as a weapon against itself. For just as there is something intellectually incoherent in relying upon an attitude of scientific tolerance in order to justify a position that is itself extremely intolerant, so too is there a problem in relying upon the reliability of our scientific methodology in order to show that our scientific methodology is in fact unreliable. If it really is the case that we cannot trust our scientific methodology, then we can hardly appeal to that very same methodology when trying to make a point about contemporary climate science. Indeed, if it really is the case that we cannot trust our scientific methodology, then we cannot really appeal to it *at all*, no matter what our intentions might be. The argument from history is therefore one that simultaneously undermines any reasons we might have for accepting its conclusion—the intellectual equivalent of sawing off the branch while you're still sitting on it. Or better yet, it is like Richard Nixon telling you that all politicians are liars, or my writing here that you can't believe anything you read in a book.

In order to better appreciate the absurdity of the argument, consider the following simplified situation. Suppose that we perform an experimental test, and come to accept the truth of one scientific theory over its various rivals. For the sake of concreteness, suppose that we attempt to measure the deflection of light as it passes through a gravitational field, and that we consequently come to accept Einstein's theory of relativity over either Newtonian Mechanics or Aristotle's theory of motion. Let us say that the experiment gives us evidence in favor of the theory of relativity. Now suppose that someone else comes along and invites us to consider the overall track record of our scientific theories. This individual points out that the majority of our scientific theories have in fact failed to predict the observed deflection of light. That is to say, both Newtonian Mechanics *and* Aristotle's theory of motion have been shown to be false. It follows then that we can extrapolate from this fact to the conclusion that all of our scientific theories are eventually going to fail—including the theory of relativity. Let us say then that the track record of our scientific theories gives us evidence against the theory of relativity. But that means that the same experiment both gives us evidence in favor of the theory of relativity (since it successfully predicted the deflection of light), and via the argument from history simultaneously gives us evidence against the theory of relativity (since most of our scientific theories failed to successfully predict the deflection of light). But one experiment cannot simultaneously provide evidence for and against the *same* scientific theory. Yet that is precisely what the argument from history entails.

Knowing the unknown

In both of the examples considered above—whether we appeal to the open-minded spirit of scientific inquiry in order to justify a point of view that brooks no opposition, or whether we extrapolate from the past in order to undermine our predictions about the future—our attempts at epistemological judo fail. It is simply foolish to try and use the strength of science against itself. More specifically, it should be clear that we cannot use the methods of science to show that the methods of science don't work.

But by the same token, it should also be clear that we can hardly use the methods of science to show that the methods of science are successful. We cannot rely upon the success of our previous scientific theories in order to argue that our contemporary scientific theories are also likely to be true. The problem here is not that any such argument undermines itself by appealing to those very factors that it seeks to discredit. Rather, it is an argument that already presupposes what it is attempting to show—and as we have already seen, this is also highly problematic. If we really did entertain any doubts about the reliability of our scientific methodology, it would not help to be given an argument that explicitly relied upon those very methods. To vary the example given above, it would be like trusting Richard Nixon just because he told us that he is not a crook.

Either way then, it would seem that we cannot offer a scientific assessment of our scientific practice. But there is a deeper reason for this difficulty. For despite initial appearances to the contrary, the argument from history is not a straightforward scientific argument. In the case of climate science, for example, it is important to note that the argument from history does not itself offer any concrete predictions about rising global temperatures or the depletion of our natural resources. Rather, it offers a prediction about how *other scientific predictions* concerning the future of our environment will fare in the future. The argument from history is therefore primarily concerned with the scientists who make the predictions, rather than the content of those predictions themselves. It is an argument about the development of society and the growth of human knowledge—and a precise prediction to the effect that our currently held convictions about the environment and our dwindling natural resources will eventually come to be seen by future generations as erroneous.

The point then is that the argument from history is not simply another scientific argument to be considered alongside our best scientific theories concerning the effects of carbon dioxide and the specific heat capacity of the polar ice caps. It is instead a *sociological prediction* about how the scientific community will come to evaluate those scientific theories in the future. And once we come to appreciate this distinction, we can better appreciate the difficulties involved in making such a prediction. For while the methods of the social sciences are closely related to the methods of the physical sciences, there are nevertheless a number of important differences between the two, and a failure to appreciate this fact can have disastrous consequences for the accuracy of the predictions we make.

Our old friend Karl Popper offered a helpful analysis of some of these differences. All scientific prediction relies upon a degree of simplification and abstraction. This is because the real world is simply too complex to understand without prioritizing certain elements, abstracting away from superfluous details, and making educated assumptions about which parts of the overall system can be safely ignored for the sake of simplicity. In particular then, a good scientific experiment is one that abstracts from the complexities of real life in order to produce a small, manageable system that can be easily subjected to repeated tests. So to take one of our favorite examples, we know that it is hopeless to try and study acceleration due to gravity by simply dropping cannonballs off the top of a tower. Conducting an experiment on such a large scale would make it extremely difficult to carry out precise measurements of the results. Moreover, any results obtained would be influenced by any number of extraneous factors, such as the imperfections in the cannonballs, variations in the way they were dropped from the tower, interested spectators interfering in the experiment, and of course the all-important influence of air resistance. In order to conduct a successful investigation, we need to isolate the relevant elements in highly controlled laboratory conditions—or even through the abstractions of pure thought in the style of Galileo.

The problem, however, is that these requirements of abstraction and simplification are especially difficult to achieve in the context of the social sciences. It is not simply that human behavior is more complex than that of a falling cannonball, although that may well be true. Rather, the problem is that human behavior is strongly determined by a vast range of intricate social interactions, and that any attempt to isolate a specific individual or group from that larger social context will *in itself* have a significant influence upon the objects of study. As Popper writes:

> Physics uses the method of experiment; that is, it introduces artificial controls, artificial isolation, and thereby ensures the reproduction of similar conditions, and the consequent production

of certain effects ... [however] artificial isolation would eliminate precisely those factors in sociology which are most important. Robinson Crusoe and his isolated individual economy can never be a valuable model of an economy whose problems arise precisely from the economic interaction of individuals and groups.[6]

It is possible to study a falling cannonball in isolation from the effects of air resistance without thereby influencing the rate of acceleration due to gravity or altering the mass of the cannonball. But when we come to study the complexities of human behavior, any attempt to eliminate some of the myriad social interactions that comprise the whole will inevitably have consequences for the rest of the system. Another way to see the problem is to note that when we measure the time taken for a cannonball to hit the ground, we can happily repeat that experiment again and again. We simply pick up the cannonball, climb back to the top of the tower, and drop it over the side. The fact that we have already performed the same experiment hundreds of times before will make absolutely no difference to the outcome of our results. Thus we can keep on studying the same cannonball in order to get ever more accurate results. But for a complex, interacting human society, the very fact that a particular experiment has already been performed can have a significant influence upon the outcome. Original participants may be older and wiser, the solution to puzzles now well known and understood, lessons have been learnt and errors corrected. More generally, the very fact that it is known that an experiment is being performed can have a significant influence upon the behavior of the subjects—individuals being willing to be on their "best behavior" in order to impress the scientists involved. In simple terms, complex social systems often exhibit a kind of feedback mechanism that is simply absent from the sort of experiments conducted in physics and astronomy.

The most important argument Popper offers, however, is also the simplest. The development of human society is strongly influenced by the state of human knowledge. It will help to determine the actions we take, and the outcomes that we expect. It will also determine the sort of technology that we develop, which in turn will have a significant impact upon the development of society. It follows then that in order to make a reasonable prediction about the development of human society, we must be able to make a reasonable prediction about the development of human knowledge. But this is simply impossible. As Popper puts it:

> if there is such a thing as growing human knowledge, then we cannot anticipate today what we shall know only tomorrow ... no scientific predictor—whether a human scientist or a calculating machine—can possibly predict, by scientific methods, its own future results.[7]

Suppose that we could predict right now the state of our future scientific knowledge. Suppose for example that we could know now that we will come to reject Einstein's theory of relativity in favor of another account of the nature of gravitational attraction and the orbit of the planets. But if we could know now that we will come to adopt such a theory, then we must presumably also be able to know the various reasons and considerations leading us to adopt such a theory. But if that was the case, then we would already know whatever evidence it was that led us to reject the theory of relativity—but if we already knew *that*, then we would have *already* rejected the theory of relativity in favor of its successor. It follows then that in order to know the future state of our scientific knowledge, we would already have to know now what we supposedly will only know in the future. The whole notion is therefore completely absurd. But if we cannot know how our future knowledge will develop, neither can we know how our human society will develop.

Environmentalism and the open society

Although it may have seemed initially plausible, the fact of the matter is that we cannot make reliable scientific predictions *about* the reliability of our scientific predictions. It is not a case of simply doing more science. Rather, when we attempt to assess the future prospects of our scientific theories, what we are really doing is making a prediction about the development of society and the growth of our scientific knowledge—a task which is not only very different from the more familiar examples of predicting the orbit of the planets or the timing of the next solar eclipse, but also one which is considerably more difficult. The argument from history with which we began this chapter can therefore be happily dismissed. Just because many scientific theories have turned out to be false in the past, that in itself does not give us good reason to suppose that our contemporary scientific theories will turn out to be false in the future. Indeed, if we can say anything with confidence about the history of science, it is that it is probably best understood as a process of gradual refinement, one whereby our scientific theories are carefully modified over the course of countless experiments and critical tests, slowly converging on an ever more accurate picture of the world. In particular then, just because so many of our previous predictions about the imminent exhaustion of our natural resources have proved to be so spectacularly wrong in the past, it does not in itself give us good reason to dismiss the latest round of apocalyptic climatic predictions. Any attempt to prophesy the evolution of human history and the longevity of our contemporary scientific theories is not only scientifically unsound, but also intellectually flawed.

Unfortunately however, the foregoing considerations cut both ways. The reason that the argument from history fails is because—as Popper puts it—we cannot anticipate today what we will only know in the future. Yet this was precisely the mistake made by William Stanley Jevons when he predicted that the United Kingdom would run out of coal by the end of the nineteenth century. What Jevons failed to take into account was the way in which improvements in our scientific knowledge would in turn influence the rate at which our coal reserves would be depleted, including the possibility of utilizing alternative sources of fuel, and the development of more efficient methods of using those resources. This is what Keynes noted when he remarked that Jevons omitted to make adequate allowance for the progress of technical methods—as well, presumably, as the declining fashion for wrapping everything up in thin brown packing paper. The fact of the matter is that when Jevons attempted to predict the rate at which the United Kingdom would exhaust her coal reserves, what he was really doing was making a prediction about the development of Victorian society and the growth of her scientific knowledge, which was every bit as misguided as our attempt to predict the future reliability of our scientific theories.

All of which leaves us with something of a dilemma. For on the one hand, the difficulties involved in predicting the way in which our scientific knowledge will evolve over time shows us that we cannot dismiss our contemporary climatic predictions simply on the basis of their poor track record. Yet on the other hand, the *reason* why we cannot draw such negative conclusions about the reliability of our environmental predictions also imposes considerable constraints on what we can expect those predictions to show. Jevons was unable to accurately predict the exhaustion of our coal reserves because he failed to take into account the increasing importance of petroleum as a source of fuel. Similarly, Paul Erhlich was utterly mistaken about the threat of mass starvation across the Indian subcontinent because he refused to take into account the ways in which improved agricultural techniques and a generally rising standard of living across the globe would affect food consumption. Edward Goldsmith was unduly optimistic about the imminent collapse of industrialization because he failed to take into account the ways in which that industrialization would evolve over the course of his lifetime. And so on and so forth for all the other environmental doomsayers we encountered earlier. Any prediction concerning the exhaustion of a natural resource, or the environmental consequences of a particular course of action, must necessarily involve a prediction about how society will respond to those eventualities—which in turn involves a prediction about the development of that society and the growth of its scientific knowledge. But since we cannot accurately predict those developments or the growth of that knowledge, it follows that we also cannot accurately predict the environmental consequences of our actions either.

None of this is of course an argument for ignoring what our contemporary scientific theories have to tell us about the environment. Nothing about the foregoing reasoning casts doubt on the fact that average global temperatures have risen slightly over the last hundred years, or that the increased presence of carbon dioxide can affect the amount of heat trapped in the atmosphere. But it does caution us against some of the more far-reaching predictions that have been made on the basis of that evidence, and which inevitably involve the rather murky business of foreseeing the development of society. In very simple terms, measuring the rise of global temperatures is science, but predicting the consequences of those temperatures—and in particular how our society will respond to them—is not. The science might be settled, but everything else remains pure speculation.

It is interesting to note the way in which this very same error of attempting to predict the growth of our scientific knowledge underlies both extremes of the contemporary climate change debate. When the "denier" dismisses any concerns we might have about the effects of industry on the environment, on the basis that such apocalyptic pronouncements have been so unreliable in the past, he is making the extremely optimistic assumption that our future scientific knowledge will show such fears to be unfounded. Conversely, when the "alarmist" announces that the extinction of the human species is imminent, on the basis that continuing our present course of action can only lead to disaster, he is making an extremely pessimistic assumption about our ability to grow and adapt. But each assumption is just as flawed as the other—for if we cannot anticipate today what we will only know in the future, then in particular we cannot anticipate if that knowledge will help or hinder us.

For Popper, this inability to accurately predict the development of society also has important consequences for how we should frame our social policies. It should caution us against the sort of large-scale top-down planning that seeks to bring about a number of far-reaching social consequences at a single bureaucratic stroke. The problem is simply that the larger the plan, and the more far-reaching the consequences, the more strongly one must assume something about the development of society and the growth of our scientific knowledge—and thus the more unreliable any such planning must become. This is in contrast to undertaking a number of smaller, bottom-up interventions that tackle individual issues one at a time. The point is not that this sort of piecemeal approach is any better able to make the relevant predictions about the development of society, but merely that by attempting to achieve less, the consequences for getting it wrong will be similarly less drastic. It is a matter of managing our mistakes, of building a policy explicitly designed to limit the extent to which our imperfect knowledge will inevitable derail the process.

The sort of sweeping reforms characterized by the comprehensive failure of international summits like the 2009 Copenhagen

Climate Conference were not only diplomatically unrealistic, they were in fact scientifically unsound—again, not in terms of the claims about industrial pollution and surface temperatures, but in the way in which these concerns were conflated with the distinct sociological question of how they should be addressed. Moreover, for Popper there is a real political danger inherent in these large-scale plans. He writes:

> The reason is that every attempt at planning on a very large scale is an undertaking which must cause considerable inconvenience to many people, to put it mildly, and over a considerable span of time. Accordingly there will always be a tendency to oppose the plan, and to complain about it. To many of these complaints the Utopian engineer [i.e. large-scale planner] will have to turn a deaf ear if he wishes to get anywhere at all; in fact, it will be part of his business to suppress unreasonable objections. But with them he must invariably suppress reasonable criticism too. And the mere fact that expressions of dissatisfaction will have to be curbed reduces even the most enthusiastic expression of satisfaction to insignificance. Thus it will be difficult to ascertain the facts, i.e. the repercussions of the plan on the individual citizen; and without these facts scientific criticism is impossible.[8]

When Popper was writing, he had in mind the sort of command economies of the Soviet Union and Communist China, which have brought such misery to their own populations and to the world at large. Both attempted to justify their respective systems through appeal to spurious scientific considerations regarding the course of history and the nature of man. These governments may have fallen, but the dangers of getting science wrong remain.

7

Deus ex machina

I think that the very first thing I was told when I began studying the history of science was that there was no such thing as the scientific revolution. It was probably a pretty dramatic opening for a lecture all things considered, although I am sorry to say that much of its intellectual impact was rather lost on me, since I had not previously been aware that there was such a thing as the scientific revolution in which I was supposed to have mistakenly believed. I was apparently so ignorant of the subject *that I wasn't even wrong* about the history of science, let alone in a position whereby I could be stimulated and challenged by the professor, and eventually led to a deeper understanding of the topic. In my case, the problem was further exacerbated by the fact that the next eight weeks of the course were subsequently devoted to outlining in meticulous detail all of the unique developments and circumstances of the period usually referred to as the scientific revolution—the mid-sixteenth and seventeenth centuries, more or less beginning with Copernicus' *De Revolutionibus Orbium Caelestium* in 1543, and reaching its culmination with Newton's *Philosophiae Naturalis Principia Mathematica* in 1687—which apparently were not all that unique or important after all. It turns out that academia is often like that: it is one thing to demonstrate how frightfully clever you are by overturning and rejecting some long-cherished traditional opinion; but sometimes you also need to spend some considerable time and energy explaining to everybody just what that long-cherished traditional opinion was before anyone will be suitably impressed that you've rejected it.

It was not until many years later that I finally grasped the point that my lecturer was trying to make. As anyone even remotely familiar with comic books and their recent cinematic manifestations can attest, one of the most natural and compelling ways we have of trying to understand the world around us is *the origin story*, an attempt to locate a specific moment of time at which everything began. We want to know how Superman arrived from Krypton, or how Batman became so psychologically scarred—or better yet, how someone with Jean-Claude Van Damme's unmistakably francophone accent could have a long-lost American brother brutally injured in a no-holds barred kickboxing competition in Thailand. If we can only pinpoint the exact circumstances in which something started, we can then understand how everything else followed, and more importantly, why things are the way they are now. Similarly, if we can properly determine when and why the scientific revolution occurred, we will also be able to better understand the various complexities of modern scientific practice. The point my professor was trying to make, however, was that it is in fact very difficult to isolate one particular moment in time, or one specific intellectual achievement, as the point at which modern science as we understand it first began. And indeed, we have already seen some of the problems involved in drawing such rigid boundaries. In the case of Copernicus, for instance, we know that his celebrated heliocentric model of the solar system was not an original innovation, but was in fact an idea that had been knocking around for hundreds of years, and had even been discussed by Ptolemy as far back as the second century. We also know that while Copernicus may have been responsible for articulating this heliocentric model with a previously unmatched level of mathematical sophistication, his motivations for doing so did not exactly distinguish him as a modern scientific thinker. Copernicus' proposal was not based upon any new experiments, did not produce any new or improved predictions, and could not even be said to be any simpler than the Ptolemaic system in any straightforward respect. The fact of the matter was that Copernicus was driven primarily by an idiosyncratic obsession with perfect geometric circles and a semi-mystical reverence for the Sun that marked him out as more of a medieval occultist than the father of modern science.

One problem then with trying to specify the origins of modern science is that human activity is never so clearly regimented and unambiguous that we are going to find an exact point in time before which everyone was desperately scrabbling around in ignorance and superstitious fantasy, and after which absolutely everything was suddenly subjected to rigorous experimentation and double-blind peer-review. Copernicus provided a detailed mathematical framework within which the geocentric model of the solar system could be challenged—but was motivated by a religious world view that Ptolemy himself would have dismissed as ridiculous. Galileo articulated a theory of inertial motion that finally made plausible the idea that the Earth too could be spinning rapidly through space—but his most spectacular and celebrated demonstrations of the new scientific method never actually happened, and he was eventually silenced by the rest of the scientific community out of a combination of jealousy and political expediency. Newton's greatest achievement was to show how both terrestrial and extraterrestrial phenomena (that is, both falling apples and orbiting planets) could be incorporated within the exact same set of physical laws, a grand unified theory of its time that irreversibly expanded the scope of our scientific horizons—yet he devoted most of his time to alchemy and esoteric speculations, and explicitly thought of gravitational attraction as God directly intervening in nature in order to hold everything together.

It therefore turns out to be quite a challenge to find some moment in time by which we would be confident in saying that the world had indeed become truly scientific. Indeed, given the wide range of pseudointellectual nonsense that still permeates our society—take homeopathy and psychoanalysis as just two of the least controversial examples—we might even begin to wonder whether there was ever a scientific revolution after all. But let us grant that there has been some development, some progress in our investigations into the hidden mysteries of nature and science. By contrast, the other problem with trying to specify the origins of modern science is in knowing how far back we need to go. We might argue with some justification that, one way or another, Copernicus was at least responsible for a whole series of significant developments in our scientific thought. So while Copernicus might have failed to fully emancipate himself from prescientific ways of thinking, it was nevertheless his work that got the ball rolling in the first place. But Copernicus was not working in a vacuum. To take just one very simple example, he had to have already been familiar with the Ptolemaic model of the solar system before it would have even occurred to him that he might be able to improve upon it and remove all of those terrible eccentrics and equants that caused him such concern. So in that sense, then, maybe we should argue instead that the scientific revolution really began with Ptolemy, who was after all ultimately responsible for motivating Copernicus. But then again, much of what Ptolemy produced was in fact a compilation and systematization of already existing astronomical observations—so perhaps we should really trace the scientific revolution all the way back to whomever first started recording the night sky. Similarly, Galileo may indeed have put together the necessary framework for finally overturning the Aristotelian principles of motion that had provided such an obstacle for accepting Copernicus' heliocentric model—but then again, it was only within the context of the Aristotelian world view that the issue of inertial motion could have become apparent and demanded a solution. So perhaps we should trace the scientific revolution back to Aristotle in the fourth century BC, who did at least start putting together a range of theories and observations against which later thinkers could rebel. But this of course simply raises another question in turn, since Aristotle himself also frequently referred back to an existing tradition and to other earlier thinkers, and so on and so forth.

At a certain point, of course, the historical record simply gives out on us. Probably the earliest historical source that we can uncover in the Western tradition is the Greek philosopher Thales of Miletus, who was working at around about the beginning of the sixth century BC, on what is now the west coast of Turkey. Thales' contribution to modern science—and in that sense, his claim to have been the ultimate originator of the scientific revolution—was the groundbreaking proposal that *everything is made of water*. Admittedly, at first sight, this may not sound like a particularly impressive scientific hypothesis, not least because it is so clearly false. But in fact the situation is even worse than this, since no one is really sure what Thales even meant by "water" in the first place, nor for that matter the sort of relationship he had in mind when he spoke of one thing being "made of" another. In fact, if we are completely honest about it, we don't really know very much about Thales at all, let alone the fine details of his scientific world view.[1] We don't know for instance whether Thales believed that everything in the universe was currently constituted by one type of substance—and that therefore macroscopic differences in the world around us could be accounted for by the different ways in which this substance was arranged—or if he believed that everything in the universe had somehow emerged or evolved from some single, original substance. As far as contenders for the most important scientific theory of all time go, the claim that everything is made of water does leave a lot to be desired.

What is particularly interesting about Thales, however, is not so much the content of his theory, but rather the sort of reasoning that it embodies. Ultimately, the central feature of Thales' account is the idea that we can attempt to explain the world around us *in its own terms*, rather than by appealing to something lying outside of the system. To say that everything is made of water could, for example, be understood as proposing a particular internal structure common to the different objects and phenomena that we encounter in the world, and that by understanding the properties and behavior of this underlying structure—how individual chunks of this "water" interact with one another—we can in turn explain the larger objects and more complex phenomena which it constitutes. Or alternatively, to say that everything is made of water could be understood as specifying some general, abstract principle that unifies the different phenomena in which we are interested. Either way, what Thales proposed is that we can look at the objects themselves in order to understand how they work. It is, in short, the idea that the world is an intelligible and self-contained system, and that scientific investigation as we know it is in fact possible as a distinct form of enquiry. And at the time, that turned out to be a very radical idea indeed.

Like the generations of leaves, the lives of mortal men

Another way to think about what is interesting about Thales' account—we might even say revolutionary—is that it is an explicit rejection of the search for origins. When Thales proposed that we could attempt to understand the world around us by looking at the internal structure of its individual parts, he was in fact rejecting a long-cherished traditional opinion that all explanations ultimately involve tracing an event back to its beginnings. This was a view that permeated almost all aspects of Thales' intellectual context, from literature and politics, to religion and history, and it was in rejecting that tradition and opening up the possibility of a new form of enquiry that his true innovation lay. Ironically enough then, the reason why Thales of Miletus has a claim to have been the ultimate source of the scientific revolution is precisely because he made respectable the view that searching for the ultimate origins of anything—the scientific revolution included—was not, in fact, a worthwhile exercise. My history professor would have approved.

It is worth exploring in a little bit more detail just how revolutionary Thales' idea—that a satisfying explanation could, in a sense, proceed *inwards* rather than attempting to understand everything in terms of their *origins*—really was at the time. We are, of course, all familiar with the dysfunctional soap opera of Greek mythology, whereby various bad-tempered and quite spectacularly promiscuous gods manipulate the natural world and its hapless mortal inhabitants, usually for lack of anything better to do. The thunder rumbles and the lightning strikes because Zeus is having a tantrum and venting his rage upon the Earth. There are no underlying mechanisms explaining the storm, nor general principles of meteorology that allows for its predictions—it can only be traced back to its source as a bad day on Mount Olympus. The sea churns and the ground trembles because Poseidon is in one of those moods. The contours of the land mark the aftermath of some cyclopean combat rather than the gradual process of erosion, the rhythm of the seasons the manifestation of some deity's ongoing domestic dispute with his wife.

And as with so much in life, this emphasis upon origins for the ancient Greeks was also closely tied up with their politics. As with many early societies, one's position in ancient Greece was largely determined by one's tribal or familial relationships, which is to say, in terms of one's genealogical origins. It followed, for instance, that the truly virtuous man did not act out of abstract ethical principles, but rather out of respect for the ties of kinship and blood; he was expected to put his family before all others and would have found the modern idea of treating all equally bordering on the incomprehensible. It is not merely for narrative purposes therefore that when the heavily armed psychopaths of Homer's *Iliad* meet one another on the beaches of Troy, every mortal combat is preceded by a lengthy recollection of the individual's family tree:

> The noble son of Hippolochus answered staunchly, "High-hearted son of Tydeus, why ask about my birth? Like the generations of leaves, the lives of mortal men. Now the wind scatters the old leaves across the earth, now the living timber bursts with the new buds and spring comes round again. And so with men: as one generation comes to life, another dies away. But about my birth, if you'd like to learn it well, first to last—though many people know it—here is my story."[2]

What follows is a lengthy tale of a mythical ancestor, deceived by a jealous queen and exiled to fight a fearsome monster, who wins fame and fortune in distant lands before being brought low by pride and the resentment of the gods, who begets a son who begets a son ... all recounted apparently in the midst of lethal hand-to-hand combat. But such recollections do not merely serve to introduce the characters and give us a sense of their prowess, like the interminable dialogue in a cheap martial-arts movie. These genealogies literally explain why such events have come to pass, since it is one's familial origins that determine one's obligations and motivations. The sons of Hippolochus and Tydeus are at Troy *because* at some point in their distant past someone swore an oath to someone else. It is an explanation that focuses upon the origins of these two men, rather than any individual calculation or personal happenstance; and like the generation of leaves, the lives of mortal men—that whatever holds true of the political world of wars and honor and debt must also hold true of the physical world of storms and thunder and lightning.

The idea then that perhaps some things are better explained in terms of their underlying structure was indeed a revolutionary development on the part of Thales. But it should also be noted immediately that Thales' contention that everything is made of water had important *narrative* advantages too. One of the problems with the broadly genealogical world view that so dominated early Greek thought is that it could often lead to conflict. The ancient Greek who understood his moral obligations in terms of his position within a familial or tribal structure, for instance, could find himself honor-bound to avenge himself against the same individual to whom he owed a personal debt. The dutiful ruler could be faced with an irresolvable dilemma between acting in the best interests of his people, and acting in the best interests of his family. The pious daughter might be forced to either disobey her father or shame her brother. And without a broader or more general ethical framework—whether that be through appeal to some higher (presumably divine) authority, utilitarian calculation, or some other moral code—such individuals literally had no way to resolve their conflict. It is this tension that forms the heart of Greek tragedy, where an individual finds themselves in dire straits, not through moral weakness or lack of virtue, but in fact as a direct consequence of faithfully fulfilling their mutually inconsistent obligations. In the theater these dilemmas were eventually resolved through divine intervention, when after much gnashing of teeth and rending of garments a god would quite literally be lowered from the rafters on the end of a pulley—the *deus ex machina*, the god from the machine—and resolves everything with a magical flick of the wrist.[3]

Indeed, it is in grappling with ethical problems of this sort that we find philosophers like Plato and Aristotle explicitly moving away from the genealogical framework. Just as Thales opened up the possibility of understanding physical phenomena in terms of a more general explanatory principle—even if we are not perhaps entirely sure what that more general principle might be—so too do we see Plato in his earlier dialogues trying to find some more general moral principle under which these conflicting obligations can be understood. We find him variously interrogating what it means to be a good father or a good king, or indeed what it means to be a virtuous daughter or an honorable guest, and reaching the conclusion that there must be something all of these have in common. In particular, he argues that in all of these cases there is a sense of having fulfilled one's obligations or duties, even if the exact nature of what these obligations entail will of course vary from case to case. What we are left with then is a very abstract ethical property—not so much of being kind to your family or ferocious to your enemies, but rather simply *doing what you should*—that provides a general organizing principle holding all of these individual instances together. In his later work, Plato would come to call this abstract property *justice*, which he identified as something like maintaining an appropriate balance between the variously competing passions of the mind. Similarly, in Aristotle's ethical work we see him arguing that the flourishing of all human life ultimately comes down to maintaining an equilibrium between the same basic, overarching goals—although with an inevitable skew towards the theoretical speculation of which Aristotle himself was particularly fond. In both cases, however, it is a moral philosophy that looks inwards at the underlying structure and raging psychologies of the subject, rather than his role in some external familial hierarchy.

And again, just as the search for origins reflected a particular social order based around ties of tribal and familial loyalties, so too did the shift towards internal structure and general principles evolve alongside a changing political climate. At the time Thales was writing, the center of power in ancient Greek society was slowly shifting from the palaces and courts of the aristocratic warlords found in the work of Homer, to the forums and marketplaces of the relatively self-contained city-state. As the social horizon extended and life began to be dominated by the intercourse of trade and commerce as much as by the Homeric strongman, the traditional framework of personal vendettas and historic grudges became increasingly disruptive to the functioning of the community. It was in this context then that philosophers like Plato and Aristotle sought to rethink the basis for their ethical interactions by abstracting to more general principles and looking inwards to the underlying psyche of their protagonists—a break from the search of origins that ultimately derived from Thales' insistence that everything is made of water.

On teleology and tragedy

To return to our main narrative, it is interesting to note how this familiar literary trope also helps us to understand some of the tensions underlying early scientific thought amongst the ancient Greeks, as well as its subsequent development. To take an obvious example, consider again Aristotle's theories of motion, and his contention that every object in the universe moves under its own peculiar inclination to return to its natural resting place, whereby stones fall downwards towards the center of the Earth, and fire burns upwards to rejoin the celestial heavens. For all of his undoubted contributions to the development of modern science, this remains ultimately a genealogical picture of the world, as there are no internal mechanisms or abstract principles determining this motion: it is simply part of what it means to be a stone that it seeks to return to its natural state of rest following whatever initial cosmic calamity that originally exiled it from its homeland. The theory is at heart a story that appeals to the goals and desires—and most importantly, the historical background—of its central characters, rather than their intrinsic properties or more general principles of motion.

But now consider one of the more problematic applications of this approach, and the way in which projectiles are supposed to behave within this Aristotelian picture. Consider for a moment a thrown stone, or an arrow loosed from a bow. Since, for Aristotle, everything moves under its own tendency to return to its natural place of rest, this kind of motion—an extended displacement in a horizontal line, rather than the stone simple plummeting directly towards the ground the moment it leaves the hand—seems particularly difficult to understand. For a projectile to keep moving in a straight line, there must be on this account some continuous force acting so as to keep it moving towards its target. But while this sort of account might make sense for the more mundane examples of a horse pulling a cart, or some purgatorial Greek hero pushing a boulder up a mountain, it is more difficult to see how such a picture applies to a projectile; after all, once the stone has left the hand, or the arrow released by the string, there does not seem to be anything still in contact with the projectile to apply the necessary force. The problem, of course, was eventually resolved by Galileo who recognized the more general principle of inertial motion—and that once the arrow is in motion, the only problem is to explain how the combination of air resistance and gravitational attraction are sufficient to eventually *make it stop*, rather than continuing forever as it would in the vacuum of space. For Aristotle, however, who could only understand motion in terms of the inclination of the object or the intervention of an outside agency, it looked as if a thrown stone should in fact plummet to the earth as soon as it was released.

Aristotle's solution to this problem—or at least, one of his solutions to this problem, as he vacillated on the matter throughout his writing—was to argue that as the projectile moved, the air that it displaced would be pushed aside and rush in behind it to fill the space left by its passage. It was this displaced air that provided the constant pressure necessary to keep the projectile in motion, slowly diminishing over time due to the counteracting effects of air resistance, at which point the projectile falls to the ground. It was not a very satisfactory solution as many of his medieval commentators noted with some consternation. It couldn't explain for instance why, if you attach tassels and streamers to an arrow and fire it into the air, the tassels are swept back in flight rather than being pushed forward as Aristotle's mechanism would suggest, or why it should be easier to throw a stone further than a ball of feathers of the same size and dimensions.

But more importantly, such reasoning comes close to inconsistency. For on the one hand, Aristotle treats the air as the principal motivating factor in the flight of a projectile—it is the compressed air in front of the arrow or stone that immediately rushes behind to fill the vacuum and thus propels the projectile forward. But on the other hand, if that is the case, we cannot *also* treat the air as some kind of impeding force that eventually overcomes the forward motion of the projectile. Either the air rushes behind the projectile to push it forward, or it stays where it is to impede its motion, but not both at the same time. So if Aristotle is right that projectiles are carried forward by the rapid backfilling of displaced air, then there is in fact no explanation as to why such projectiles should ever stop moving at all. By thinking of the motion of objects in genealogical terms, as something ultimately determined by the agency of the objects and their ultimate origins, we seem to reach contradictory conclusions about the motion of projectiles. They must either never move at all, since their own natural inclination will take over immediately once they are released; or they must move eternally, since the only factors capable of slowing them down are completely devoted to keeping it moving forward.

Yet despite such undoubted difficulties, the genealogical approach remained stubbornly persistent. One explanation for this of course is simply the natural inertia of an existing intellectual tradition—which is particularly ironic when one considers that it is in fact the very notion of inertia that the Aristotelian framework has such difficulties accommodating. But there is also a more general difficulty. It may indeed be tempting to try and explain the behavior of individual objects through an investigation of their underlying structure, but it is not always so clear as to how we might extend such theorizing to the world as a whole. The problem, of course, is that the world as a whole is *completely unique* insofar as there are no other worlds with which it can be compared. It follows then that there can be no general abstract principles under which it can be grouped, no broader patterns of behavior to which we can appeal. Similarly, it is difficult to see what might count as the underlying structure of the world as a whole, since any smaller component—individual objects, subatomic particles, whatever it was that Thales had in mind when he spoke about "water"—will simply be part of the very world that we are trying to explain.

And once we begin upon such a line of thought, the question naturally arises as to how any of Thales' protoscientific explanations could even be possible in the first place. The very fact that certain parts of the world can be made sense of in terms of some general principles or underlying structure seems *itself* to call out for further explanation, which takes us right back to the question of origins all over again. Both Plato and Aristotle asked how we could not suppose the world to have been fashioned by some powerful creator or "demiurge"; for a more modern example, the English naturalist William Paley asked:

> In crossing a heath, suppose I pitched my foot against a stone, and were asked how the stone came to be there: I might possibly answer, that, for any thing I knew to the contrary, it had lain there for ever; nor would it perhaps be very easy to show the absurdity of this answer. But suppose I had found a watch upon the ground, and it should be inquired how the watch happened to be in that place; I should hardly think of the answer which I had before given — that, for any thing I knew, the watch might have always been there ... [W]hen we come to inspect the watch, we perceive (what we could not discover in the stone) that its several parts are framed and put together for a purpose, e.g. that they are so formed and

adjusted as to produce motion, and that motion so regulated as to point out the hour of the day; that, if the different parts had been differently shaped from what they are, of a different size from what they are, or placed after any other manner, or in any other order, than that in which they are placed, either no motion at all would have been carried on in the machine, or none which would have answered the use that is now served by it.[4]

It is the flip side of Thales' approach that while it may allow us to transcend some of the contradictions inherent in the narrower genealogical approach, it must also presuppose certain limitations to what it can explain. When looking for origins we can naturally go beyond the system in question; but in focusing upon what can be explained *within* that system, we cannot also call the system itself into question.

A planet wholly inhabited by spiders

In a chapter devoted largely to the narrative of science, it is perhaps only to be expected that our middle act should end in a twist. I had been trying to suggest that one of the most useful ways to think about the origins of science lies not in any particular discovery or experimental result, but rather in a shift in the sort of story that we tell about the world around us. In particular, I have been trying to suggest that a scientific explanation—thought about in very abstract terms—is an attempt to make sense of things in terms of themselves, their internal structure, or their organizing principles, rather than seeking to trace their origins to some external (often divine) source. Somewhat ironically, I have tried to make this case in part by tracing back some of the origins of modern science, but let us put this down to dramatic flair.

We have seen a number of advantages to this approach, and the way in which the break from genealogical thinking was reflected in similar developments in the arrangement of moral and political life in ancient Greece. Nevertheless, we have also seen that there are some substantial limitations to this sort of approach, which surprisingly enough seem to bring us back right to where we started. It is one thing to explain individual events and phenomena in terms of their internal structure, or some higher organizing principle—but when dealing with the world as a whole, there seems to be no additional internal structure to which we can appeal, and certainly no higher organizing principles for what is by definition a completely unique object. When it comes to making sense of *this*, the entire mystery of existence for want of a better term, we seemed left with no other option than to fall back on a search for origins. In trying to make a break from genealogical reasoning, we find ourselves forced back upon it in order to explain why other sorts of explanation may be possible. In trying to trace the origins of scientific thinking, we find ourselves again making an appeal to the divine.

Traditionally, there have been a number of different arguments advanced that seek to establish the existence of God, or at least some kind of all-powerful creator for the world. Some of these arguments have attempted to make the case on purely logical grounds, painstakingly unraveling concepts and very piously chopping up definitions, while others have simply appealed to the strength of personal conviction in the individual in question. Neither approach has proved particularly compelling for those not already largely convinced of the conclusion. But what the above reasoning suggests is something like a more empirical approach. In order to follow Thales in breaking with the traditional search for origins, and to frame our explanations in terms of the internal structure or higher organizing principles of the phenomena in question, we find ourselves faced with the question as to why the world should be so arranged that there would be the right sort of internal structure or higher organizing principles that make such explanations possible. It raises the question as to why the world should exhibit such clear design or purpose in its construction—and suggests that the only explanation must be to trace the origins of the world back to some initial act of creation. This line of thought is known as the *teleological argument* for the existence of God, coming from the Greek word *telos* meaning design or purpose, and it appears to be a consequence of what we had been exploring as some of the distinctive features of scientific thought.

Nowadays, of course, it is widely supposed that this sort of teleological reasoning about the world has been decisively refuted by the development of evolutionary biology—or at least, given the continued popularity of intelligent design theory amongst certain school boards in the United States, that such reasoning *should* have been decisively refuted by the development of evolutionary biology. It is certainly agreed that the conjunction of genetic mutation and natural selection provides us with an alternative explanation of the apparent evidence of design, even if there perhaps remains some disagreement as to whether or not it provides us with a *better* explanation of that evidence. Moreover, it also provides us with the sort of explanation of which Thales of Miletus would have approved, in that it seeks to give us an account of some of the most general features of the world in terms of the interaction of earlier and simpler parts, rather than by appealing to some further, external source. In that sense then, evolution seems to help us realize the scientific ideal of Thales, an intellectual framework that resolves the lingering doubts of those like Plato, Aristotle, or Paley in showing how the genealogical break could be completed. Indeed, given the manifold difficulties in characterizing genuine scientific activity in terms of its falsifiability or (even worse) its methodological naturalism, this broadly *narrative* contrast between what we might think of as the creationist origins story, and the evolutionary emphasis on intrinsic properties, may in fact offer a more plausible criterion of demarcation.

But while evolutionary biology has undoubtedly become the focus for contemporary opposition to teleological reasoning, there are a number of serious problems with this way of understanding the situation. The twists just keep on coming. In providing an alternative explanation for the apparent well-ordering of the natural world, the theory of evolution certainly defeats any presumption that divine intervention is the only possible way for such features to arise. Even the most committed of creationists has to admit that the argument from design falls somewhat short of the logical guarantee of a mathematical proof. But the same is true of any other alternative explanation that we might be able to come up with, not just the specific account inspired by Darwin—and just as we can trace examples of the design argument going all the way back to antiquity, so too can we establish an equally impressive pedigree for its opposition.

Writing as early as the 1750s, roughly 100 years or so before the publication of Darwin's *On the Origin of Species*, our old friend David Hume mused on the various shortcomings of teleological reasoning. Like Galileo before him, Hume presents his potentially problematic theological views in the form of a dialogue between three friends, rather than a straightforward philosophical treatise. The discussion remains superficially pious throughout, concerning itself not with the existence of God as such, but rather with the different ways in which men might come to know of his existence—although the attentive reader cannot miss the fact that all such methods are found to be irreparably flawed by the end of the conversation. In the dialogue, Cleanthes represents the view of the natural theologian who maintains that God is revealed through his creation, and duly presents a variety of teleological arguments to that effect. He is opposed in turn by Demea, a more old-school theologian who believes that knowledge of God comes through faith and rational reflection, and who eventually leaves the conversation in somewhat of a huff; and Philo, a general skeptic and troublemaker who largely represents Hume himself.

One of the principal problems identified by Philo is that the design argument is highly subjective. There are lots of different ways that we might interpret the evidence around us, and therefore lots of different potential explanations that we might offer. In particular, even if we suppose that complex biological phenomena could not have come into existence through random chance, it seems peculiarly self-important to suppose that the ultimate cause must resemble some kind of intelligent human design. As Philo rather amusingly puts it:

> The *Brahmins* assert, that the world arose from an infinite spider, who spun this whole complicated mass from his bowels, and annihilates afterwards the whole or any part of it, by absorbing it again, and resolving it into his own essence. Here is a species of cosmogony, which appears to us ridiculous; because a spider is a little contemptible animal, whose operations we are never likely to take for a model of the whole universe. But still here is a new species of analogy, even in our globe. And were there a planet, wholly inhabited by spiders, (which is very possible), this

inference would there appear as a natural and irrefragable as that which in our planet ascribes the origin of all things to design and intelligence, as explained by *Cleanthes*. Why an orderly system may not be spun from the belly as well as from the brain, it will be difficult for him to give a satisfactory reason.[5]

I am not familiar with the elements of Hindu cosmogony to which Hume here refers; but I take it that none of his readers would have missed the elegant insinuation that the advocate of the design argument is—just like the spider—pulling it out of his arse.

Later in the dialogue, Hume even considers the possibility of explaining the apparent evidence of design in terms that are more statistical than supernatural, and whether or not well-adapted organisms might have arisen at random through something approaching a process of trial and error. Again, expressing the idea through the skeptic Philo:

> It is in vain, therefore, to insist upon the use of the parts in animals or vegetables and their curious adjustment to each other. I would fain know how an animal could subsist, unless its parts were so adjusted? Do we not find, that it immediately perishes whenever this adjustment ceases, and that its matter corrupting tries some new form? It happens, indeed, the parts of the world are so well adjusted, that some regular form immediately lays claim to this corrupted matter: And if it were not so, could the world subsist? Must it not dissolve as well as the animal, and pass through new positions and situations; till in great, but finite succession, it falls at last into the present or some such order?[6]

Of course it would be completely anachronistic to suggest that Hume here anticipates the Darwinian revolution in biology. He does not, for instance, suggest any kind of mechanism whereby this process of trial and error might occur. But in terms of the philosophical evaluation of teleological reasoning, he does capture in its entirety the relevance of evolution—that well-ordered organisms can be found in the world around us, not because some higher intelligence so contrived to put them there, but simply because any less well-ordered organisms would not have survived long enough for us to encounter them.

But that is not how the narrative is supposed to go. It was supposed to be that the Darwinian revolution finally provided us with the intellectual resources to overcome this kind of teleological reasoning, to complete the break with genealogical thinking that went back to Thales. There are certainly plenty of academics and popular writers today who take that line, and have made a career out of publishing books arguing that case. But it is difficult to see how that could be true if the essential ideas were already in circulation a good century before Darwin set sail for the Galapagos Islands; the details of his theory may well have helped to flesh out and make more vivid this alternative, but ultimately added absolutely nothing of content to the issue.

So to begin with, we have a search for the origins of modern science that paradoxically locates it in the rejection for any search for origins. At the same time, however, in exploring the limits of this intellectual shift, we find ourselves faced with yet another search for origins, this time regarding the very existence of a world sufficiently well ordered to allow different types of nongenealogical explanation. And the usual way in which we expect to resolve this conflict—through the development of a broadly evolutionary style of thinking that replaces intentional design with the relentless trial and error of natural selection—we find to be completely irrelevant to the issue at hand. At this point our narrative may have strayed into the operatic rather than the Homeric, and it is time to bring it all together.

The big finale

It is perhaps then not so surprising that genealogical thinking about the natural world has managed to survive the Darwinian revolution. After all, the central explanatory ideas underlying the theory were already well known and in circulation for almost a century before *On the Origin of Species* even hit the shelves, and had done very little to dent the enthusiasm for explaining the world around us in terms of its initial origins—indeed, one of the most famous examples of this sort of genealogical thinking at the hands of William Paley had actually been proposed many years after Hume's full-frontal assault. Another iteration of the same explanatory strategy was therefore never going to settle the issue once and for all, even if the superior technical articulation of Darwin's theory of evolution by natural selection did of course prove more persuasive than the ideas nascent in Hume's more conversational presentation. The problem is simply that, while the principles of natural selection certainly offer an alternative account of the existence of biological complexity than one which simply traces their origins to some divine craftsman, this is not itself sufficient to show that the genealogical approach must therefore be false. Even if we accept that the principles of natural selection offer an unequivocally *better* account of the existence of biological complexity, this is also not sufficient to prove that William Paley was wrong, since sadly the best explanation is not always the right explanation. Waxing lyrical about the undoubted merits of evolutionary thinking has an important rhetorical function, but it can never make any progress against the underlying logical point.

Hume, of course, was well aware of the philosophical limitations of such an argumentative strategy, which is why even though he does indeed present an absolute barrage of alternatives to the design argument, this is not in fact the main thrust of his attack. Hume's central contention is in fact that there is a serious tension at the heart of such genealogical reasoning. This is perhaps easiest to appreciate once we realize that the arguments developed by Plato, Paley, and indeed contemporary creationists, are all an *argument from analogy*. The starting point is relatively uncontroversial. We all agree naturally enough that when we do encounter an artefact or machine deliberately designed for a purpose—such as a pocket watch left lying on the heath—it is perfectly reasonable to infer that there must be someone originally responsible for its creation. The argument then proceeds by inviting us to consider the analogous degree of order and purpose in the natural world, and then to conclude that as a matter of intellectual consistency we must also infer an analogous creator for the world as a whole.

The problem, however, is that the analogy has to satisfy two competing demands. For on the one hand, we want the argument to be as *compelling* as possible, and this requires that the analogy between material artefacts and well-adapted species be as close as possible. The more that we can convince ourselves that pocket watches and opposable thumbs must indeed be considered as displaying the same sort of deliberate design, the more plausible it will be to conclude that both require a similar sort of explanation. On the other hand, however, we also want the conclusion of the argument to be as *complete* as possible, since at the end of the day the argument is not just supposed to help us conclude the existence of any old creator, but rather help us establish the existence of God. But if that is the case, then we do not want the analogy between material artefacts and well-adapted species being too close, or else we would just end up inferring the existence of a creator not all that dissimilar to ourselves—a rather lackluster deity whose abilities barely outstrip those of the local artisan, and who would certainly not justify much in the way of religious devotion and praise. These two desiderata naturally pull against one another, and present something of a dilemma at the heart of any such genealogical reasoning. Hume puts the point very succinctly when he (or rather, Philo) observes that:

> All the new discoveries in astronomy, which prove the immense grandeur and magnificence of the works of nature, are so many additional arguments for a deity, according to the true system of theism: But according to your hypothesis of experimental theism they become so many objections, by removing the effect still farther from all resemblance to the effects of human art and contrivance.[7]

The more impressive the creation, the more powerful the creator—but at the same time, the more magnificent the natural world, the less persuasive the comparison with the mundane artefacts of human craftsmanship, and the less compelling the inference that the world itself must have been similarly designed.

The real problem with the design argument therefore is not its lack of scientific credentials (we have already been down that route), nor the significant limitations of its research framework (although that is also important). The real problem with the design argument is *narrative*—it is committed to telling a particular type of story that is ultimately too linear to resolve its central contradictions. It is committed to telling a story that wants to emphasize the genealogical origins of the natural world in order to motivate the belief in an ultimate creator, yet which must also downplay those genealogical origins in order to avoid motivating the belief in the *wrong sort* of ultimate creator. It is all a bit of a tragedy, only in this case there is little hope for divine intervention to help us resolve the internal contradictions since, ironically enough, it was precisely the existence of divine intervention that the whole argument was supposed to establish.

The sort of story that we end up telling as part of our scientific investigations can therefore be almost as important as our methodology or choice of experiments. But just as overemphasis upon any particular aspect of our methodology can often lead us astray when thinking about science—such as reducing all aspects of science to the process of falsification, or exalting the role of experiment without recognizing the importance of interpretation—so too can an excessive focus upon any particular narrative style also raise difficulties. It is one thing to abandon a broadly genealogical approach to the complexity of the natural world, and the idea that everything must be explained in terms of its ultimate origins; but this is not the same thing as rejecting the need for explanation altogether, and care must be taken lest the former leads to the latter.

The underlying narrative tension in the design argument concerns how exactly we interpret the complexity of the natural world, that on the one hand we need to think of it as relevantly similar to human craftsmanship in order to strengthen the analogy, while on the other hand we need to think of it as significantly surpassing human craftsmanship in order to motive a suitably divine origin. One of the lessons that we can draw from this realization is that the complexity of the natural world is always going to be a largely *subjective* matter—that the mathematical harmony of the heavens or the surprising adaptation of a species to their environment all comes down to your point of view. And from there it is not a big step to come to the opinion that in fact the world around us does not seem to be particularly well ordered or designed at all. The unnecessary death and needless pain, ruined hopes and shattered dreams, and the vast reaches of desolate and empty space. Looked at from a certain angle, we might even reach the conclusion that the whole of creation is just one almighty accident, hostile and meaningless, a compelling argument *against* the existence of any kind of divine creator who would not possibly have allowed such a state of affairs to come to pass.[8]

Such a view has become increasingly popular in our increasingly secular society, and indeed comes with its own peculiar narrative tradition, perhaps most notably in the work of the existentialists in the early half of the twentieth century, and continuing through the variously nihilist strands that we might loosely classify today as postmodernist. In these works—in novels and plays as well as the more straightforwardly academic publications—it is a commonplace to encounter the fundamental *disorder* of the world in what is often an amusing inversion of the sort of divine revelation usually associated with the design argument. In his novel *La Nausée*, Jean-Paul Sartre gives us a rather prosaic example—his protagonist, Antoine Roquentin, a struggling writer with what appears to be an increasingly loose grasp on the world around him, sits in a park and is struck by the overwhelming pointlessness of it all. Reflecting upon his experience, Roquentin tries to put it into words:

> Comic ... No: it didn't go as far as that, nothing that exists can be comic; it was like a vague, almost imperceptible analogy with certain vaudeville situations. We were a heap of existents inconvenienced, embarrassed by ourselves, we hadn't the slightest reason for being there, any of us, each existent, embarrassed, vaguely ill at ease, felt superfluous in relation to the others. *Superfluous*: that was the only connexion I could establish between those trees, those gates, those pebbles ... I dreamed vaguely of killing myself, to destroy at least one of these superfluous existences. But my death itself would have been superfluous. Superfluous, my corpse, my blood on these pebbles, between these plants, in the depths of this charming park. And the decomposed flesh would have been superfluous in the earth which would have received it, and my bones, finally, cleaned, stripped, neat and clean as teeth, would also have been superfluous; I was superfluous for all time.[9]

Which can sound somewhat depressing, but is in fact a perfectly enjoyable notion to ruminate upon in the warmth of a Parisian cafe with a couple of stiff drinks inside you.

Yet while such intellectual posturing may have much to recommend it to a secular and disaffected age, it is a universal acid that must be handled with care. A world devoid of all order and structure is not only a world without a divine creator, it is also a world that must forever elude our capacity to predict and explain. Without some underlying principle of organization, it remains just as arbitrary and superfluous to see the world in terms of individual objects operating under the laws of physics as it does to see it in terms of divine craftsmanship. In order to do science at all, we have to assume some kind of ordering principle in the great cosmic chaos in which we find ourselves—perhaps not another God from the Machine, but certainly enough to pause for thought.

Epilogue

The philosophical study of the natural sciences only really began to take off at the end of the eighteenth century. Before that point of course philosophers had harangued one another and argued interminably over such pressing issues as the fundamental nature of reality and whether or not everything we experience is all just a dream. But the particular questions with which this book has been concerned—the objectivity of scientific practice, the distinction between genuine scientific research and pseudoscientific nonsense, and what we even mean by the scientific method—are, in fact, a relatively recent innovation. They were largely inspired by the work of Isaac Newton, who on the one hand greatly simplified our scientific understanding of the world by showing how both terrestrial and celestial phenomena could in fact be described by the same basic principles of mechanics, while on the other hand greatly complicated matters by couching those principles in such advanced mathematical language that few people could readily understand them. Naturally enough this raised some serious concerns as to what exactly a scientific theory was supposed to achieve, and whether or not being able to *understand* the world around us was still the same thing as being able to accurately predict its behavior. Moreover, as our scientific theories gradually revealed a world increasingly at odds with our everyday experiences—a world displaced from the center of the universe and hurtling around the Sun at unimaginable speeds despite all apparent evidence to the contrary—it became all the more pressing to explain how such counterintuitive images could nevertheless provide us with accurate and reliable knowledge.

This problem exercised the minds of some of the greatest thinkers of the time, but it was the German philosopher Immanuel Kant who was to have the most significant influence over the debate. The central idea underlying Kant's philosophy was the claim that knowledge is an essentially collaborative enterprise—he argued that while the external world may provide us with raw data through the medium of our senses, it must also be shaped and organized by our cognitive faculties before it can constitute meaningful information. This was in contrast to a broad philosophical consensus at the time that saw learning as a purely passive activity, one where we begin life equipped with the mental equivalent of a blank canvas upon which we patiently wait for the external world to leave its mark. As Kant pointed out, however, without some kind of organizing principle this would produce little more than a confusing jumble of sound and fury that would make absolutely no sense at all. Kant devoted a great deal of his philosophical activity to investigating how exactly our cognitive faculties shape and organize our knowledge, and argued that certain abstract principles—such as the idea that one and the same physical object can persist in time, or that one event invariably follows another—are precisely the sort of organizing principles that we supply to our experiences in order to construct an intelligible world.

More importantly though, it followed that if all of our knowledge is shaped and organized by our own cognitive faculties, there must be some very general things that we can know with absolute certainty simply by reflecting upon the way in which those faculties operate. It stands to reason that if you go out in the morning wearing rose-colored spectacles, then you can be sure that everything you see that day will have a pinkish hue. Similarly, if all of our experience has to be carefully packaged into distinct spatial chunks persisting through time before we can understand it, then we can know for instance that the principles of geometry must hold true for all times and places—again, not because they are something that we can read off the external world after lengthy and detailed investigation, but merely because they describe some of our preconditions for encountering the external world in the first place. And what goes for mathematics and geometry also goes for the natural sciences. According to Kant, the reason why we can be confident of the complex and counterintuitive claims of our scientific theories is because the fundamental principles of Newtonian Mechanics—that all motion will continue in a straight line, that every action has an equal and opposite reaction—are themselves simply expressions of the way in which we shape and organize our experience of the world.

Kant's solution to the problem was ingenious and technically accomplished, but it also had another important motivation. As well as attempting to place developments in the natural sciences on a more secure intellectual footing, there were broader social and political challenges that Kant's philosophy was designed to address. For while Newton's achievements had without doubt ushered in a new spirit of optimism regarding mankind's ability to make progress through reason and rationality alone, they had also put considerable pressure on the traditional principles of morality and religion that held society together. This was partly a question of political legitimacy, since if every man is held to be equally capable of exercising his own reason, it was no longer clear why he should continue to defer to the established authorities of Church and State. But it was also partly a question of social order, for in a world entirely governed by invariable physical laws, where each and every action was fully determined by underlying mechanical principles and man was seen as little more than a complex machine, there seemed to be no room left for either free will or genuine moral agency. Thus, as the advance of our scientific knowledge held out the prospect for a rational solution to the problems of society, it simultaneously undermined the political institutions and moral framework upon which that society was based—a tension that would culminate in the guillotine of the French Revolution and the bloody upheavals that followed across Europe, and which in many ways we are still wrestling with today.

From its very earliest inception therefore, the philosophy of science can be seen as an explicitly political enterprise, as Kant attempted to negotiate between two competing goals: to provide a satisfying intellectual justification for Newtonian Mechanics, while at the same time carefully delineating the scope of human reason so as to leave room for the more intangible virtues of faith, hope, and charity. So on the one hand, Kant attempted to show how we could know some things with absolute certainty, on the grounds that they merely described the contribution made by our own cognitive faculties in shaping our experience; while on the other hand, he also attempted to show that there were some things that were in principle *unknowable* by the exercise of pure reason alone, and which included the realm of ethics and religion. In particular, Kant drew a sharp distinction between the everyday world of our experience—shaped by our own cognitive faculties and therefore constructed in absolute conformity to the principles of arithmetic, geometry, and Newtonian Mechanics—and the world lying behind those experiences, the raw material out of which we construct our everyday lives and whose intrinsic nature must forever remain beyond our grasp.

It was to prove a delicate balancing act that produced a number of internal schisms amongst those who followed Kant, and which still define the major antagonisms and competing factions of modern philosophy today. Nevertheless, Kant's general framework for understanding and justifying our scientific theories was to remain dominant for another century of solid philosophical bickering. In essence, the idea was that we could trust our scientific theories—no matter how exotic or mathematically inaccessible—because just as with the principles of geometry and arithmetic, they were in fact an expression of the fundamental workings of the human mind. It was therefore seen as quite a blow when mathematicians began exploring the possibility of different systems of geometry at the end of the nineteenth century, and an absolute disaster when Einstein overthrew Newtonian Mechanics completely at the beginning of the twentieth century. If our scientific theories merely describe the fundamental structure of the human mind, it was difficult to see how we could even seriously consider rival candidates for the job, let alone change our minds altogether and replace one scientific theory with another.

Philosophers are, however, nothing if not resilient, and they were never going to let something like a revolutionary development in the natural sciences derail their long-cherished opinions about how it was supposed to work. An urgent rescue operation was duly initiated. It was argued that while we are somehow responsible for organizing and systematizing the raw data of our experience as Kant maintained, the mechanism by which we do so must be considerably more flexible than the unchanging structure of our cognitive faculties. The problem was that Kant had pitched the level of his analysis far too high by attempting to provide an analysis of *reason in general*, rather than the specific scientific theories that are its result. But this is a difficult task, since good reasoning does not tend to consist of a fixed set of propositions that apply for all times and circumstances, but is often crucially determined by context. Inevitably then, what Kant provided was a philosophically high-tech canonization of the common sense wisdom of his day, and as Einstein showed, what is considered common sense today may not be considered common sense tomorrow.

It was proposed instead, therefore, that what Kant had envisaged as the way in which our cognitive faculties shape their incoming experience was rather more like a process of laying down formal definitions in a language—we stipulate what we mean by terms like "time" and "space," general axioms and principles for determining their use, and let these guide the way in which we experience the world. Thus, it was that one of the principal achievements of Einstein's work was not in accumulating more empirical data or constructing ever more rigorous experimental tests, but rather in offering a new way of understanding what it meant for two events to be "simultaneous" with one another, and from which it was possible to provide a cleaner way of accommodating our existing principles of mechanics with electromagnetism. By the 1920s and the 1930s therefore, the entire philosophy of science had become almost a branch of theoretical linguistics, concerned not with deducing the fundamental principles of human cognition, but with the painstaking definition of scientific terms, clarification of the language of our scientific theories, and with tracing the purely semantic links between the two.

And again, this development went hand-in-hand with an explicit political agenda. In the period immediately following the horrors of the First World War, there was a fervent hope that the calm rationality of science would offer a much-needed corrective to the jingoistic nationalism and diplomatic brinkmanship that had plunged Europe into a bloodbath of unprecedented proportions. But it was not just a matter of attitude—science also held out the prospect of a true universalism, a common language and set of principles that could transcend the narrow tribal interests that had caused so much suffering. By clarifying our scientific vocabulary and clearly delineating the relationships between our different scientific concepts, these philosophers of science were not merely attempting to reformulate Kant's general project, they were explicitly engaged in an effort to widen its democratic participation. Complex theoretical language was to be reduced to simpler statements about observational consequences, and the underlying relationships made explicit in an unambiguous logical framework. By making science accessible to the masses, it was hoped that they could be provided with the tools and information necessary to actively engage with the political process, which in turn would lead to a more equitable and just society.

Inevitably of course, such utopian thinking eventually led to its own demise. By explicitly associating our scientific theories with something as arbitrary as a language, the so-called linguistic turn in philosophy made it irresistible to thereby associate our scientific theories with a *culture*. In both Nazi Germany and the Soviet Union, for instance, Einstein's theory of relativity was summarily dismissed as "Jewish Science," nothing more than a particularly insidious way of talking about the world rather than the outcome of scientific rigor and experimentations—and today, in humanities departments across the world, you will still find groundbreaking scientific achievements "problematized" for their role in perpetuating forms of colonial oppression. Just as we might worry about imposing our own moral values and political institutions upon unwilling participants in the imperialist enterprise, so too has the critique of primitive superstition and inferior levels of technology become seen as an objectionable instance of cultural chauvinism. Ironically enough then, the net result of this project has been the abandonment of scientific universalism in favor of a dogmatic form of multiculturalism that emphasizes and thereby cements the differences between people, and which subsequently breeds precisely the kind of social divisiveness it was supposed to remedy.

Philosophically speaking, the entire approach was also bedeviled with technical shortcomings. For while Kant's approach was ultimately too rigid to accommodate the reality of scientific change and innovation, the linguistic turn was by contrast too flexible. If our experience of the world really is constituted in part by the linguistic framework we choose to apply, then there must be a sense in which *different* linguistic frameworks provide *different* answers to the same scientific questions. The idea of having conflicting scientific theories would therefore be equivalent to the idea of mutually untranslatable languages, ones where central terms such as "space" and "time" in Newtonian Mechanics literally have no equivalents in a relativistic vocabulary. But the notion is incoherent, since in order to even recognize something as another language presupposes that we can make enough sense of it to see it for what it is, which in turn means that we can see how certain words and phrases are supposed to refer to objects and events, which in itself constitutes a rudimentary translation. In order to be sure that we really did have an example of an untranslatable language, we would have to explain—in our own language—which parts of the other language could not be translated. The very fact that we will necessarily always be able to translate between these different linguistic frameworks shows that they are not in genuine competition—which means that there can be no sense in which they really do give different answers to the same scientific questions. In the end, therefore, a linguistic framework is just *too* flexible for the job at hand.

The philosophical end result has been the eventual abandonment of Kant's search for certainty. There are no logical guarantees that our scientific theories are accurate and reliable, since they do not reflect either the inner workings of our minds nor our deliberate set of linguistic stipulations. And this has been in many ways epistemologically liberating—we can acknowledge that our scientific theories give us our best means for finding out about the world around us, but without the unrealistic expectation that their results will be forever beyond reasonable doubt. The fact that our scientific theories evolve and develop means that no analysis of our scientific knowledge can ever be considered complete, but will only ever be provisional on our currently accepted scientific world view. This is a consequence that most philosophers of science will happily concede, since they would also maintain that our epistemological investigations are continuous with our best scientific practices, and that there is no higher perspective from which we can approach these questions. We are to undertake what has been championed as a thoroughly *scientific* investigation of our scientific practices.

But it is here that contemporary philosophy of science has found itself run aground. Our scientific theories are themselves part of the natural world, an intellectual tool for interacting with our environment just like a sharpened stick or a sundial, and therefore themselves the subject of scientific investigation. We are then left with the vicious circle of having to appeal to our most trusted scientific theories in order to help us determine which scientific theories we have good reasons to trust. The result has been philosophy reduced to its most basic elements, with those predisposed to trust science finding themselves in a position to offer scientific arguments in support of their claim, while those of a naturally more skeptical view of the scientific process are similarly able to marshal the same amount of scientific support in favor of their own position. It is a philosophical argument that only produces whatever you put into it, and differences of opinion, so often masked in academic dispute behind sophisticated argument and counterargument, are laid bare as unadulterated intellectual prejudices. And this opens the way to the use of science as a political tool—no longer rooted in the fundamental structure of the human mind or a rigorous set of definitions, and neither tasked with delineating the space of reason nor providing the basis for a universal discourse, science threatens to become little more than an expression of one's deeper and more personal convictions. That is a dangerous development, and it has been the subject of this book.

Dramatis personae

Aristotle (384–322 BC)

Greek philosopher and polymath whose surviving work covers a dizzying array of topics from physics, biology, metaphysics, formal logic, art, poetry, theater—and, of course, the obligatory studies in rhetoric, politics, and practical governance. In contrast to his teacher Plato, who emphasized the values of theoretical speculation and saw the deductive certainty of geometry and mathematics as the ideal forms of human enquiry, Aristotle was concerned with the ways in which we gradually build up our body of knowledge from the repeated observation of specific instances, and is often credited with establishing the methods of modern science. Aristotle's views on matter and motion dominated scientific thought until the sixteenth century, when advances in astronomy began to put pressure on his essentially static view of the universe. Later in life, Aristotle served as a tutor to the young Alexander the Great, and partly inspired his extraordinary eastward conquests through a steady diet of unquestioned Greek cultural supremacy and an unfortunately flawed grasp of global cartography.

Claudius Ptolemy (c. 100–170)

Greek mathematician, astronomer, and geographer, Ptolemy lived as a Roman Citizen in the city of Alexandria in Egypt. His *Almagest* is the oldest surviving treatise on astronomy, and its geocentric model of the universe remained the authoritative text well into the sixteenth century when it was eventually overturned by the heliocentrism of Copernicus. Ptolemy also wrote an important cartographical treatise that collated the existing geographical knowledge of the Roman and Persian Empires, and other studies in astrology, optics, and music.

Like many ancient authors, Ptolemy's work was lost from Europe following the collapse of the Roman Empire, and only recovered during the Renaissance from Arabic sources.

Nicolaus Copernicus (1473–1543)

Polish mathematician and astronomer, and somewhat unwilling participant in the endless political machinations between the Prussian Empire and the Monastic State of the Teutonic Knights, Copernicus is of course best known for his heliocentric model of the universe (*De Revolutionibus Orbium Coelestium*). While not the first astronomer to suggest that the Earth orbited the Sun, Copernicus was the first to articulate the idea with sufficient mathematical rigor, and his work is widely taken to mark the beginning of the scientific revolution in Europe. Copernicus was initially unwilling to publish his work, and when it finally came to print under the supervision of the Lutheran theologian Andreas Osiander, a disclaimer was inserted stating that the proposed heliocentrism need not be taken to be "true or even probable" in order to generate accurate predictions—thus establishing an entire branch of the philosophy of science to which some of us have devoted entire academic careers.

Galileo Galilei (1564–1642)

Italian mathematician, astronomer, and physicist, Galileo developed the theories of mechanics necessary for accommodating a heliocentric understanding of the universe, i.e. a rapidly spinning Earth, with our everyday experience of a largely static firmament. While Galileo's work was initially welcomed by the Catholic Church, subsequent political maneuvering soon found Galileo in trouble. His work was banned, and Galileo lived out his days under house arrest in Florence. The Inquisitions's ban on reprinting Galileo's work was finally lifted in 1718, and the general prohibition on books advocating heliocentrism removed in 1758. Nevertheless, the issue was revived in the early nineteenth century by Protestant polemicists keen to paint an image of the Catholic Church as a reactionary and dogmatic institution. The fact that it was the Protestant Church that first raised concerns regarding the heliocentrism of Copernicus—and which has continued to be the first to condemn other scientific advances apparently in conflict with scripture—was conveniently forgotten.

Isaac Newton (1642–1726)

English mathematician, alchemist, physicist, self-experimenter, and heretic, Newton's monumental *Philosophiae Naturalis Principia Mathematica* in many ways marked the culmination of the scientific revolution begun by Copernicus and Galileo, showing how terrestrial and celestial mechanics could be reduced to the same basic principles of motion—a system of such mathematical elegance that it reigned unchallenged for nearly 300 years, and is still endorsed as a limiting case of the relativistic mechanics of Einstein. The archetype of the idiosyncratic genius, Newton's appointment to the Lucasian Professor of Mathematics at the University of Cambridge was almost scuppered by his refusal to renounce (or indeed, hide) his extreme religious unorthodoxy, eventually requiring a direct royal intervention by Charles II. Promoted to an honorary position as Master of the Royal Mint, Newton personally went undercover in some of the worst bars and taverns in London to track down counterfeiters; while as President of the Royal Society, he blatantly abused his position to attack and discredit his own rivals in a vicious priority dispute over the invention of calculus. Newton's recorded time as a Member of Parliament, however, seems to be confined to a single intervention—when he complained of a draught and asked that someone close the window.

David Hume (1711–76)

Arguably one of the greatest philosophers of all time, Hume was never able to secure a proper university position, and eventually had to resort to politics to make his living, working first as the Secretary to the British Embassy in Paris, and later as the Under Secretary of State for the Northern Department. Hume's philosophy is built around the notion that all knowledge must come ultimately from experience—and that this, in fact, provides a very meager sustenance. Thus, Hume argues that most of our convictions concerning the external world and the regularity of nature are based on little more than habit and expectation rather than rational inference (*An Enquiry Concerning Human Understanding*); and that our moral principles reflect more our turbulent emotions than they do our dispassionate reason (*An Enquiry Concerning the Principles of Morals*). This extremely skeptical perspective underlies Hume's particularly notorious argument against the existence of miracles: he argued that since any miracle will be an extraordinary and highly unlikely event, it is in fact always more probable that our eyes are deceiving us or that we are being deliberately deceived than it is that the miracle actually took place; thus any evidence for any religion is by definition untrustworthy. And this is precisely the sort of smart-ass comment that explains why Hume always had such bad relationships with his academic colleagues

Immanuel Kant (1724–1804)

Born in the city of Königsberg in Prussia (now the city of Kaliningrad in Russia), where he spent almost the entirety of his life, Kant's work marks something of a watershed in the development of modern philosophy. In what he termed his own Copernican Revolution, Kant argued that the objects of our knowledge must in part be shaped and determined by our own cognitive faculties in order to make sense—in opposition to the then-dominant view that experience is a purely passive activity. However, while Kant devoted considerable effort and technical sophistication as to how this process might actually work, he was never able to fully satisfy all of his critics. It is only a slight exaggeration to say that the current schism that exists between the so-called continental philosophy favored in Europe, and the contrasting analytic philosophy of the Anglo-American world, derives from the respective preferences for either the first or second edition of Kant's *Critique of Pure Reason*. Famous for his otherwise rather uneventful life, popular myth maintains that Kant never travelled more than 16km from Königsberg throughout his entire life, and that he showed a remarkable lack of interest in the outside world—although as scholars have been quick to point out, he did briefly work as a tutor 20km away in the small town of Veselovka, and once asked a friend for news from Berlin.

Charles Darwin (1809–82)

English naturalist and biologist, whose *On the Origin of Species* introduced the general public to the principles of evolution and natural selection. Darwin proposed, and outlined in fantastic detail, how variation between species could arise through a gradual process of adaptation to different environments, and his book is widely credited as laying the foundations of modern biology. However, the publication of the work was somewhat fraught, Darwin having originally begun to formulate his ideas in 1836 following a research expedition to the Galápagos Islands on *HMS Beagle*, but finally only motivated to put his ideas in writing in 1858 when it looked as if he might be beaten to the punch by Alfred Russell Wallace. There remains much scholarly interest into why Darwin took so long to publish his work, including fear of hostility from the ecclesiastical authorities, and a difficult family life punctuated with the illness of his children. For anyone who has spent considerable time amongst serious academics, however, the delay is somewhat less surprising.

James Clerk Maxwell (1831–79)

Scottish-born physicist and experimentalist, who held positions in Aberdeen and King's College London before taking up the first Cavendish Professorship at the University of Cambridge in 1871 and overseeing the construction of the Cavendish Laboratory. Maxwell's most important scientific contributions were in the field of electromagnetism, where he provided a mathematically sophisticated extension of Michael Faraday's work on electricity and magnetism in terms of a single electromagnetic field. On the basis of further calculations, he later conjectured that light was also an electromagnetic wave, and posited the existence of an all-encompassing luminiferous aether—eventually rejected by later physicists—as the medium in which these waves were transmitted. Maxwell also made important contributions to the statistical understanding of thermodynamics, whereby the heat of a system is understood in terms of the average distribution and kinetic energy of its component particles.

Henri Poincaré (1854–1912)

French mathematician, physicist, and philosopher who made seminal contributions to an extraordinary range of topics, including number theory and topology, and helped to lay the foundations for many modern branches of study such as chaos theory and quantum mechanics. He is credited by Einstein as having made important contributions to the development of the theory of relativity, although the two men never agreed on its overall interpretation. In his philosophical work, Poincaré asked how we could trust the claims of our contemporary scientific theories when the history of science demonstrated such a degree of revision and error (*Science and Hypothesis*); his solution was to note that while many of the superficial details of our scientific theories change over time, there is nevertheless considerable continuity and progress at the level of the underlying mathematical framework.

Albert Einstein (1879–1955)

German-born physicist, Einstein renounced his citizenship in 1933 when the Nazis came to power, and after several years of uncertain status he eventually became a U.S. citizen in 1940, where he remained at the Princeton Institute of Advanced Study with many other of the greatest minds who had fled Europe. Best known for his general and special theories of relativity, Einstein also helped to establish the field of quantum mechanics—although he was never satisfied with the essentially probabilistic nature of the theory, famously declaring that "God does not play dice." Despite his firm pacifism, Einstein played an instrumental role in convincing President Roosevelt to begin research into the development of nuclear weapons (something made possible by his own scientific breakthroughs) as a precaution against the Nazis, a decision that he regretted for the rest of his life.

Karl Popper (1902–94)

Austrian-born philosopher of science who emigrated from Europe in the 1930s, first to the University of New Zealand, and after the war, to a professorship at the London School of Economics. Best known for his work on the scientific method, which he argued should be understood as the continuous process of falsification rather than confirmation (*The Logic of Scientific Discovery*), Popper's thought was in many ways simply a continuation of his political philosophy. Thus the notion that scientific practice can be distinguished from other forms of human activity by its emphasis upon critical testing evolved from his criticism of the intellectual emptiness of socialism and other totalitarian ideologies (*The Poverty of Historicism*); and his insistence that our scientific understanding is always provisional and open to revision was developed in opposition to the view that society can be effectively planned by self-appointed experts (*The Open Society and Its Enemies*). While Popper remains one of the most well-known philosophers of science of the twentieth century, his abrasive personality and lack of technical sophistication means that his work remains firmly out of favor amongst the academic community.

Carl Gustav Hempel (1905–97)

German logician and philosopher, who emigrated to the United States in 1937 in the wake of growing anti-Semitism throughout Europe, and subsequently held positions at several universities including Yale University, Princeton University, and the University of Pittsburgh. Hempel is perhaps best known for his work on explanation, which he believed consisted in the attempt to show how the event we wish to understand was in fact an inevitable consequence of the laws of nature (in conjunction with the relevant initial conditions). This general program of providing a precise, logical structure for different aspects of scientific practice placed Hempel firmly within a school of thought known variously as logical positivism or logical empiricism, and which dominated the philosophy of science in the first-half of the twentieth century.

Thomas Kuhn (1922–96)

American physicist, historian, and philosopher of science, who held professorships at University of California at Berkeley, Princeton University, and the Massachusetts Institute of Technology. His principal works include a study of astronomy in the early modern period (*The Copernican Revolution*), the development of quantum mechanics (*Black-Body Theory and the Quantum Discontinuity*), and the nature of the scientific method (*The Structure of Scientific Revolutions*). Kuhn was particularly concerned to challenge the idea that science progresses through the application of rigorous rules and principles, but is rather guided by a vague and largely subjective appreciation of shared problems and techniques that defies a more concrete articulation. Kuhn's work has often been taken to demonstrate that scientific practice is more influenced by personal interest than empirical evidence, and is often just a tool for political oppression—an interpretation that Kuhn strenuously denied throughout the rest of his career.

Jean-Claude Van Damme (1960–)

Belgian-born martial artist and movie star (born Jean-Claude Van Varenberg). Best known for his flashy high-kicking fighting style, and his signature move of doing the splits and punching an assailant in the groin, Van Damme has also explored a number of philosophical issues throughout the course of his work, and has shown a particular interest in the postmodern deconstruction of identity. This is developed through the playful absurdity of all his characters having a pronounced francophone background—"*what accent?*"—regardless of their personal circumstances, a surprisingly large number of films where he plays identical twins separated at birth (*Kickboxer*, *Double Impact*, *Maximum Risk*), and of course the questions of personal responsibility, memory, and history explored through the context of a reanimated corpse in the superb *Universal Soldier* franchise (although not *Universal Soldier: The Return*, which was terrible).

Notes

Chapter 1

1 Karl Popper, *The Logic of Scientific Discovery* (Routledge, 1959), 30. In what is probably the most well-known and influential account of the scientific method written in the twentieth century, Popper argues that the myriad complexities of the natural sciences can all be reduced to the basic underlying principles of testing and falsification. It is a book that has been praised for its elegance of thought, its logical clarity, and perhaps above all, for its readability—a rare enough virtue in professional philosophy to deserve special mention. One suspects, however, that most of these devotees have never made it past the handful of introductory sections that they pretended to read in college, since the most salient quality of Popper's work is in fact its hubris. *The Logic of Scientific Discovery* is a lengthy book, and after briefly presenting its central thesis, is primarily devoted to the highly technical debugging and refinement of this supposedly simple idea in the face of a seemingly endless succession of counterexamples and complications. It climaxes with a spectacularly inept attempt to provide a metric for comparing the content of rival scientific theories—a technical device upon which the entire logical structure of the book depends—which was conclusively refuted a few years later, and which Popper was to spend the rest of his career variously attempting to refine, reformulate, or simply trying to forget. Suffice it to say, Popper did not always practice what he preached.

2 It is important to note that the more apparently austere and traditional one's preferred flavor of religion, the less likely one is to endorse a literal interpretation of scripture—which is why you only tend to find creationism endorsed by the most modern and evangelical sects of Christianity. There is of course a historically complex explanation for this fact, stretching back through the Protestant Reformation and including such philosophically nuanced issues as the nature of divine revelation and man's relationship to God. Ultimately, however, the issue comes down to one of theological bureaucracy. To put it in very simple terms, if the Bible is to be understood as a literally true account of (amongst other things)

the ultimate origins of the world and the spontaneous creation of man, then there is no longer any need to rely upon the sophisticated interpretation and voluminous commentary provided by the countless saints and innumerable early church fathers so revered by the more conservative denominations. Creationism is therefore in fact a relatively recent innovation, resulting more from the explicit rejection of the great raft of ecclesiastical intermediaries that had somehow managed to locate themselves between the parishioner and his God than the preservation of a more venerable intellectual tradition.

3   *McLean v. Arkansas*, IV(C). Parts of the transcripts of the case can be found in various nooks and crannies of the internet, usually accompanied by the kind of vituperative invective that one tends to find when the rival sects of science and religion engage one another online. A far more edifying use of one's time can be devoted to reading Robert Pennock and Michael Ruse (eds), *But Is It Science? The Philosophical Question in the Creation/Evolution Controversy* (Prometheus Books, 2009), which collects together transcripts from *McLean v. Arkansas*, as well as a number of excellent essays on the historical and philosophical background to the controversy, useful introductions to the technical and legal issues involved, and some first-class philosophical analysis of the whole debacle. I particularly recommend the contribution from Larry Laudan, with whom I basically agree about everything.

4   Karl Popper, *Conjecture and Refutation* (Routledge, 1963), 35. On the whole, Popper's later writings are a considerable improvement over his earlier work, and begin to articulate a much more nuanced account of scientific practice. Popper's political views, including his extended attacks on Marxism and other brands of totalitarianism, are collected in the short and somewhat dry *The Poverty of Historicism* (Routledge, 1957), and the wide-ranging and highly enjoyable *The Open Society and Its Enemies* (Routledge, 1945). Highlights include the claim that Plato—the man who supposedly invented the notion of open-minded philosophical inquiry in the West—was a close-minded fascist, and that Hegel—long championed by continental philosophers as one of the foremost authority on human freedom—was an obsequious apologist for the highly repressive Prussian State. Somewhat surprisingly, Marx himself is treated with genuine respect, although his later followers are summarily dismissed with utter contempt.

5   Karl Popper, 'Intellectual Autobiography,' in P. A. Schlipp (ed.) *The Philosophy of Karl Popper* (Open Court Publishing, 1974), 137. This book contains a collection of essays on Popper's philosophy, followed by a rebuttal from the man himself, and provides a

fascinating insight into the reception of his work amongst the academic community. The general impression given is that pretty much everyone thinks that Popper is wrong, both in terms of his general approach and in his mishandling of the technical details. I do not think that there is a single aspect of Popper's thought that is not subject to the most withering assault. This of course raises the question as to how such an academically maligned set of views could nevertheless become so popular amongst the wider reading public—although on the other hand, one might suspect that it is *precisely* this popularity amongst the wider reading public that explains the degree of academic hostility that Popper encountered. The purely disinterested pursuit of truth can be a bit like that sometimes.

6 *Kitzmiller v. Dover*, 64. There are a lot of resources available relating to the trial. For an intriguing account, see M. Chapman, *40 Days and 40 Nights: Darwin, Intelligent Design, God, Oxycontin, and Other Oddities on Trial in Pennsylvania* (Harper Perennial, 2008). The title is taken from a joke made during the trial, which in fact lasted exactly forty days and forty nights. When asked by the defense if this had been deliberate, the judge replied that it "was not by design."

Chapter 2

1 Giorgio Coresio, *Operetta intorno al Galleggiare de Corpi Solidi* (Florence, 1612). For an interesting survey of some of the contemporary literature to Galileo's supposed experiment—and the almost complete ignorance of it ever taking place—see Lane Cooper, *Aristotle, Galileo and the Tower of Pisa* (Cornell University Press, 1935). Cooper was inspired to make his study upon realizing that while most of his scientific colleagues were firmly convinced that Galileo had indeed demonstrated the superiority of independent observation and experimentation over the uncritical acceptance of received authority and tradition, ironically enough none of them had in fact ever attempted to perform the same experiment themselves.

2 Vincenzo Renieri, *Letter to Galileo* (Pisa, March 13, 1641). Note the reference to "a certain Jesuit." Our good friends from the Society of Jesus will have an important role to play later in our story.

3 It is notable in that regard then that when Galileo does come to formulate what we might think of as his official opposition to the Aristotelian consensus, it is conducted in the form of a *thought-experiment* rather than an empirical demonstration. Suppose

NOTES

that we have two cannonballs of equal size and weight falling side-by-side from the top of some useful tower, which even the Aristotelian would agree will fall at the same rate. Now suppose the two cannonballs to be connected by a thin, delicate thread. Again, there is no reason to suppose that this will have any influence on the rate of descent of the two cannonballs. We now gradually shorten the length of the thread until the cannonballs are lightly touching. At this point we are now effectively dealing with one cannonball of twice the original weight — but it seems absurd to suppose that it will now suddenly double its speed.

4 William James, *The Principles of Psychology* (Harvard University Press, 1890). This is often referred to in the philosophical literature as the theory-ladenness of observation—an idea famously expounded in N. R. Hanson's *Patterns of Discovery* (Cambridge University Press, 1958), subsequently developed in the work of Thomas Kuhn and Paul Feyerabend in the 1960s and 1970s, and eventually taken to unintentionally hilarious extremes by some of the more intellectually facile schools of postmodernist science studies that unfortunately still continue to discredit humanities departments in universities around the world.

5 Considerations of simplicity and mathematical elegance were not necessarily the most important motivations for Copernicus. Waxing lyrical about his heliocentric cosmology in the introduction to his book, he writes:

> In the middle of all sits Sun enthroned. In this most beautiful temple could we place this luminary in any better position from which he can illuminate the whole at once? He is rightly called the Lamp, the Mind, the Ruler of the Universe; Hermes Trismegistus names him the Visible God, Sophocles' Electra calls him the All-Seeing. So the Sun sits as upon a royal throne ruling his children the planets which circle round him. (*De Revolutionibus Orbium Caelestium*, [§10])

As the reference to Hermes Trismegistus suggests—an individual who regular occurs in works on alchemy and other pseudoscientific esoterica—Copernicus was as much concerned with articulating his mystical vision of the world, and his quasi-divine veneration of the Sun, as he was in any of the more traditional scientific endeavors of simplifying or systematizing the existing data. In many ways then, Copernicus was in fact considerably less progressive in his thinking than many of his Ptolemaic rivals. For more on the background to this episode, see Thomas Kuhn's *The Copernican Revolution* (Harvard University Press, 1957).

6 Galileo Galilei, *Dialogue Concerning the Two Chief World Systems* (Florence, 1632), The Second Day; translated by Stillman Drake (Random House, 2001), 146. The dialogue form may well have traced its inspiration back to the philosophical work of Plato, but it also served two other important purposes for Galileo. First, it allowed him to present his ideas in a more readily accessible format for the general public—it is important to note that the *Dialogue* was also originally published in colloquial Italian rather than Latin. And second, it allowed Galileo to keep a critical distance from the Copernican ideas that he was advancing. In terms of the former, this strategy proved to be highly successful; in terms of the latter, somewhat less so.

7 Another example is the description in Ecclesiastics 1:5 of how "the Sun also ariseth, and the Sun goeth down, and hasteth to his place where he arose." Similarly, the anger of the Lord in Job 9:6-7—that "which shaketh the Earth out of her place, and the pillars therefore tremble; which commandeth the Sun, and it riseth not; and sealed up the stars"—presumably requires that the Earth is usually at rest (in order for it to be shaken), and that the Sun is usually in motion (in order for it to be commanded otherwise).

8 St. Augustine, *The Literal Meaning of Genesis* (c. 415 AD), Section 1.18.37. St Augustine was born in the Roman province of Hippo Regius in modern day Algiers, where he also served as bishop. One of the most celebrated early Fathers of the Church, St. Augustine can also boast to be the patron saint of brewers, printers, and theologians—an unusual combination to be sure, but one which pretty much covers all of my interests.

9 Or at least that's my interpretation. But I am at least in good company here, for I generally agree with Stillman Drake, one of the great authorities on Galileo, and who presents his case in the wonderfully readable *Galileo* (Oxford University Press, 1980).

Chapter 3

1 Having spent the bulk of his *Dialogue Concerning the Two Chief World Systems* explaining how the rapid motion of the Earth around the Sun would not have any noticeable effect on those clinging to its surface—since everything else around them would also be moving at the exact same speed—Galileo nevertheless proposes that it must be the rapid motion of the Earth around the Sun that is responsible for the great tidal sloshing back and forth of the world's oceans (translation by Stillman Drake, 2001, 483–4). On the face of it, this is extremely puzzling, for if a rapidly spinning Earth need not entail a constant rush of wind to the face as we struggle about our daily business, why then should it entail a constant rush of water up and down the beach as we try to build a sandcastle? It remains a matter of some scholarly dispute as to whether or not Galileo was aware of this apparent contradiction, but either way it was Newton who finally provided the systematic resolution that he lacked.

2 "An Experiment to Put Pressure on the Eye" (Cambridge University Library, Department of Manuscripts and University Archives, The Portsmouth Collection, Ms. Add. 3995, 15). Newton's notebooks from this period are a fascinating combination of his work on optics, his laundry list, the refraction of light through a prism, grisly self-experimentation and various outstanding debts. With the exception of the groundbreaking contributions to natural science of course, this reminds me a little of my time at Cambridge too.

3 A good example of the sort of the excitement and hype surrounding the development of Big Data Analytics is Chris Anderson, "The End of Theory: The Data Deluge Makes the Scientific Method Obsolete" (*Wired*, June 23, 2008), which cheerfully predicts the replacement of experiment and theory with nothing more than industrial scale number-crunching. For the original article discussing Google Flu Trends, see Jeremy Ginsberg et al., "Detecting Influenza Epidemics Using Search Engine Query Data" (*Nature* 457, February 19, 2009).

4 The reader will not be surprised to know that the obsession amongst philosophers with ravens—and what color they may or may not be—has produced a substantial literature known as the Raven Paradox. Introduced by Carl Hempel in the 1940s, this concerns a curious contradiction that seems to underlie our intuitive understanding of evidence and confirmation. It seems natural enough to suppose that if we want to test the hypothesis that all ravens are black, we would take the presence of a black raven to provide positive support for the hypothesis, the presence of a white raven as falsifying the hypothesis, and the presence of anything that was not a raven—such as a white tennis shoe, a blue coffee cup, or a red herring—as being utterly irrelevant. By the same token, if we wanted to test the slightly more contrived hypothesis that all nonblack things were nonravens, we would suppose that white tennis shoes and blue coffee cups provided positive evidence for the hypothesis (they are neither black, nor ravens), and any color of raven to be completely irrelevant. The problem is that the two hypotheses are *logically equivalent:* to say that all ravens are black *is just to say that* anything that is not black cannot also be a raven. But now we have argued ourselves into the position of accepting

that the very same piece of evidence—be it a black raven or a white tennis shoe—is both relevant and irrelevant to the same scientific hypothesis, depending upon something as insignificant as how we express it. See C. G. Hempel, *Philosophy of Natural Science* (Prentice-Hall, 1966) for the classic introduction to this problem, and pretty much any philosophy of science journal of the last sixty years for endless disagreements about how to resolve it.

5 David Hume, *Enquiries Concerning Human Understanding and Concerning the Principles of Morals* (London, 1748) Part IV; edited by L. A. Selby-Bigge (Oxford University Press, 1975), 35–6. Much of Hume's philosophical work was devoted to exposing the flaws and weaknesses in the grand intellectual theories of his contemporaries; like Galileo before him, it did little to improve his popularity.

6 Ibid., 63. It is interesting to note that while a scientist like Newton usually compared the workings of the universe to the complex mechanisms of a watch, a philosopher like Hume always seemed more readily inclined to compare it with the decidedly more human intrigues of a billiards table. The difference was probably ultimately one of personalities. Newton of course believed that everything had been carefully arranged by an all-powerful deity, and was obsessed with discovering how it all fitted together; Hume by contrast was a notorious atheist and *bon vivant* who accepted the randomness of existence with pragmatism and good cheer. Nowadays philosophers of science like to talk about billiard balls almost as much as they like to talk about black ravens, yet they somehow lack Hume's light-heartedness.

7 I take the useful distinction between sympathetic and homeopathic magic from my extremely well-thumbed copy of James Frazer, *The Golden Bough* (London, 1913). By the time of the medieval period, the magical taxonomy had become increasingly complex in order to differentiate between those sources of magic that were inherent in the world—and which may therefore be benign depending upon how they were used—and those derived from the summoning of demons and other kinds of satanic pacts, and a whole industry quickly developed in order to legislate and punish on the basis of these highly abstract considerations.

Chapter 4

1 Arthur Conan Doyle, *A Scandal in Bohemia*, first published in the *Strand Magazine*, June 25, 1891.

2   The story unfortunately does not have a happy ending, since while those institutions that copied the Vienna General Hospital by initiating a comprehensive hand-washing policy consequently saw a rapid and pronounced reduction of mortality rates, Semmelweis was nevertheless ridiculed for his suggestion, and found that his employment opportunities gradually diminished. Returning to Hungary, he became increasingly obsessed with the issue, and began writing a number of vicious open letters accusing prominent members of the medical profession of deliberately murdering their patients through their crass stupidity. When this surprisingly enough didn't work, Semmelweis turned to drink instead. In 1865, he was institutionalized. He was severely beaten by orderlies after trying to escape, and ironically enough died a few months later of an infected wound. For more on the philosophical significance of Semmelweis' case, see—again—C. G. Hempel, *Philosophy of Natural Science* (Prentice-Hall, 1966).

3   J. J. C. Smart, *Philosophy and Scientific Realism* (London, 1963). A pioneer in several areas of ethics and metaphysics, Smart is also known for a peculiar argumentative strategy for unbalancing and defeating an opponent whereby you point out the absurd consequences of their view, and then wholeheartedly embrace them as a virtue; a form of epistemological chicken, this is known colloquially in philosophical circles as "out-Smarting" someone.

4   Hilary Putnam, "What is Mathematical Truth?," *Philosophical Papers Vol. 1: Mathematics, Matter and Method* (Cambridge, 1975), 73. Many of these issues are still the topic of lively research amongst philosophers of science today, and naturally enough can swiftly become fairly complex and technical. For a more indepth discussion of the particular issue of how we should go about evaluating the reliability of our scientific theories, as well as the intellectual background to the debate, the interested reader could always consult—cough—my *A Critical Introduction to Scientific Realism* (Bloomsbury, 2016).

5   Bas van Fraassen, *The Scientific Image* (Oxford, 1980), 40. In the course of his work, van Fraassen defends the somewhat idiosyncratic view that the aim of science is in fact merely to provide us with accurate knowledge and predictions about the observable phenomena — that is, those things that we can observe unaided with the human eye—and that therefore everything else our scientific theories tell us about, e.g. microscopic entities, subatomic particles and other things that we *cannot* directly observe, is essentially irrelevant to this underlying purpose. This view is motivated by a combination of general epistemic caution regarding the scope and accuracy of our scientific theories, and a certain

pragmatism about what it is that we think science can achieve. I actually wrote my doctoral dissertation on van Fraassen's work, but don't worry, I will spare you the details.

6 Probably the most important work in this field was conducted by Amos Tversky and Daniel Kahneman, for which the latter was awarded the Nobel Prize in Economics in 2002 (Tversky unfortunately died in 1996). If you have not already done so, see Kahneman's *Thinking, Fast and Slow* (Penguin, 2011).

7 There are roughly speaking two different ways in which our cognitive faculties can go astray. If we suppose that the rustle in the grass is a tiger when there is in fact nothing there, we commit a *false positive* since we have falsely supposed that there is actually a tiger present. Conversely, if we come face-to-face with an actual tiger, but falsely suppose that we are hallucinating or that someone is playing a trick on us, then we commit a *false negative*. Given that no cognitive process is going to be 100 percent accurate, the suggestion is that the rough and tumble of evolution will generally have favored those that err on the side of committing too many false positives, since one false negative is all it takes to remove oneself from the gene pool altogether.

8 Letter from Charles Darwin to William Graham, July 3, 1881.

9 Alvin Plantinga, *Where the Conflict Really Lies: Science, Religion, and Naturalism* (Oxford University Press, 2011).

Chapter 5

1 Letter from Albert Einstein to Marcel Grossman, September 12, 1920.

2 I am indebted here to the research of Andreas Kleinert, whose talk I was fortunate enough to attend in Munich several years ago. A full version of Professor Kleinert's research was published as "Paul Weyland, der Berliner Einstein—Töter" ("Paul Weyland, the Berlin Einstein-Killer") in H. Albrecht (ed.), *Naturwissenschaft und Technic in der Geschichte, 25 Jahre Lehrstuhl für Geschichte der Naturwissenschaft und Technik am Historischen Institut der Universität Stuttgart* (Stuttgart: Verlag für Geschichte der Naturwissenschaft und Technik, 1993), 198–232. For those interested, Weyland's novel was titled *Hie Kreuz – Hie Triglaff* (*The Cross Against the Triglaff*) the latter being a pagan symbol of the aforementioned bloodthirsty Slavs who are righteously slaughtered by the heroic German knights. Unfortunately, I have not been able to find a copy on Amazon, and so cannot tell you if it is any good.

3 As Bertrand Russell put it:

> If we trace any Indo-European language back far enough, we arrive hypothetically (at any rate according to some authorities) at the stage when language consisted only of the roots out of which subsequent words have grown. How these roots acquired their meanings is not known, but a conventional origin is clearly just as mythical as the social contract by which Hobbes and Rousseau supposed civil government to have been established. We can hardly suppose a parliament of hitherto speechless elders meeting together and agreeing to call a cow a cow and a wolf a wolf. The association of words with their meanings must have grown up by some natural process, though at present the nature of the process is unknown.

Bertrand Russell, *The Analysis of Mind* (George Allen and Unwin, 1921), Lecture X: Words and Meaning.

4 Ludwig Wittgenstein, *Philosophical Investigations* (Blackwell, 1953). As with much of Wittgenstein's work, however, the exact interpretation of what he wrote remains a matter of some controversy and has divided the academic community into mutually antagonistic factions all claiming him as their own—which I guess only really proves his point.

5 Thomas Kuhn, *The Structure of Scientific Revolutions* (University of Chicago Press, 1962), 10. Kuhn was by training a historian of science who felt that his colleagues in the philosophy of science had been led astray in their more theoretical reflections by an implausible grasp of their subject matter; his book was therefore intended as a contribution to philosophy from the perspective of an enthusiastic fellow traveler. Such interdisciplinarity is often fraught with difficulties, and Kuhn's work was roundly criticized by philosophers for its lack of theoretical sophistication, generally snubbed by historians for its lack of descriptive detail, but eagerly adopted by a whole range of sociologists and other postmodern scholars for whom lack of detail or sophistication has never been considered an obstacle.

6 Ibid., 94.

7 This exchange is recounted in Plato's *Theaetetus* (c. 380 BC). The Sophists were a class of professional rhetoricians who traveled Ancient Greece offering tuition to the wealthy and privileged. Much of the education, however, tended to involve simply memorizing lengthy speeches on worthy subjects, which could then be reproduced *ad nauseam* at fashionable dinner parties—or even the public forum—without the trouble of having to form one's own opinion on the subject. I will allow the reader to draw their own comparisons here with the state of higher education today. Many of Plato's Dialogues revolve around Socrates encountering one of these highly educated youths in the market place, and after flattering them for their advanced learning, exposes both the hollowness of their understanding, and the facile political correctness of their opinions, through a process of disingenuously simple questioning. The Sophists were unsurprisingly irked by this, and were instrumental in bringing the criminal charge of "corrupting the youth" against Socrates which eventually led to his execution. This period in history is known as the Golden Age of Athenian Democracy.

8 Donald Davidson, "On the Very Idea of a Conceptual Scheme," *Proceedings and Addresses of the American Philosophical Association* 47 (1973): 5–20.

Chapter 6

1 William Stanley Jevons, *The Coal Question* (London: Macmillan and Co., 1865), 154. Like many of the environmental and economic doomsayers that have followed him, Jevons' work was explicitly inspired by Thomas Malthus's 1798 *Essay on the Principle of Population*, which famously argued that mankind would quickly outgrow its agricultural capacity to be fed—a situation only remedied by the salutary influence of war and famine. A puritanical churchman with an ecclesiastical horror for all pleasures of the flesh, Malthus was thus an early exponent of the view that most of the world's problems come down to the fact that while there are just enough of *us*, there are far too many of *them*. As is often the case, Malthus' patrician disdain for the irredeemably fecund working classes blinded him to the possibility of any kind of improvement to their situation, and when free-market capitalism offered the obvious solution through the repeal of the Corn Laws in 1846 some other mechanism was urgently required in order to reassure everyone that society nevertheless remained doomed. This was the motivation for Jevons' study, who extended Malthus' framework by arguing that "the momentous repeal of the Corn Laws throws us from corn upon coal" and that any improvements in the average standard of living amongst the lower classes must be reversed before it was too late.

2 John Maynard Keynes, *Essays in Biography* (Horizon Press, 1951), 266. While praising the brilliant and engaging writing of *The Coal*

Question, Keynes nevertheless concludes that "its prophecies have not been fulfilled, the arguments on which they were based are unsound, and re-read today it appears over-strained and exaggerated," which is about as unequivocally damning as you can get really. Keynes was a bit like that.

3 If you must, see Paul R. Ehrlich, *The Population Bomb* (Sierra Club, 1968); Edward Goldsmith and Robert Allen, *A Blueprint for Survival* (Ecosystems Ltd, 1972); Barbara Ward and René Dubos, *Only One Earth: The Care and Maintenance of a Small Planet* (W. W. Norton & Co., 1983); and Michael Oppenheimer, *Dead Heat* (St. Martin's Press, 1990)—although I wouldn't really recommend it.

4 Henri Poincaré, *Science and Hypothesis* (Walter Scott Publishing Co., 1905), 160. Poincaré subsequently goes on to dismiss this pessimistic assessment on the grounds that it misunderstands the nature of a scientific theory. For Poincaré, the purpose of science is to provide a precise mathematical description of the world around us, and while subsequent theories may supersede one another in terms of their descriptive claims—particles or waves, crystalline spheres or the deformation of spacetime—the underlying equations associated with these claims nevertheless demonstrate a high degree of continuity. Despite surface appearances, therefore, the track record of science is actually one of slow and steady progress. While many of the historical details of Poincaré's claim here remain controversial, it is nevertheless part of a distinguished line of philosophical world views that seeks to mitigate the fallibility of our day-to-day knowledge in terms of a set of deeper and more profound truths only accessible to the professional academic.

5 There is—as always—an interesting political parallel to what superficially appears to be merely an issue of scientific methodology. This is what Popper refers to as the *paradox of tolerance*, the idea that "if we extend unlimited tolerance even to those who are intolerant, if we are not prepared to defend a tolerant society against the onslaught of the intolerant, then the tolerant will be destroyed, and tolerance with them" (see his *The Open Society and Its Enemies*, Ch. 7, note 4). Popper concludes that in order to preserve a tolerant society, we must reserve the right not to tolerate the intolerant—by force if necessary—in the same way that we must reserve the right not to tolerate murder, enslavement, and other criminal activity. Similarly, it can be no part of the free and open-minded spirit of investigation to endorse a blinkered dogmatism.

6 Karl Popper, *The Poverty of Historicism* (Routledge & Kegan Paul, 1957), 8. Like much of Popper's work in the philosophy of science,

the real target of his attack remains the sort of totalitarian socialist politics from which he fled in Europe. While perhaps best known for challenging the pseudoscientific credentials of an ideology whose theories are completely unfalsifiable, Popper is also at pains to stress the intractable complexities of making predictions about the evolution of human society, and how many of the great "historical narratives" that claim to have uncovered the underlying forces shaping our destiny are guilty of this mistake.

7 Ibid., p. vi.

8 Ibid., p. 89. Popper's concerns here can be usefully compared with those of his friend and colleague Friedrich Hayek, whose *The Road to Serfdom* (University of Chicago Press, 1944) approaches the issue from a more straightforwardly political perspective. While Hayek is of course better known for his work in economics, Hayek and Popper refer to each other's work frequently in their respective writing, and develop similar concerns as to how widespread misunderstandings of the scientific method have been used to justify iniquitous political ideologues and badly flawed social policies.

Chapter 7

1 Most of what we know about Thales comes from a handful of second-hand anecdotes recorded by Aristotle, where he is variously attributed the usual list of semimythical philosophical accomplishments such as predicting eclipses, inventing geometry, and of course traveling to Egypt to study ancient and esoteric secrets. In one of his more down-to-earth exploits, however, Thales simply uses his astronomical knowledge to predict a bumper olive harvest and quietly goes about cornering the market in olive-presses, which he then rents out at exorbitant prices to all those who had previously mocked the value of philosophy. See for example Aristotle's *Metaphysics* (983b27–33), and *Politics* (1259a).

2 Homer, *The Iliad* (c. 800 BC), 6: 169–78; translation by Robert Fagles. In this particular instance, the genealogical interlude proves somewhat useful, as the two warriors discover that their grandfathers were, in fact, close friends, and therefore decide upon closer reflection not to kill each other after all, exchanging armor like competitors from rival sports teams. In Homer, the Trojan War lasts for ten years, which may seem slightly less surprising if we are to suppose that such episodes are truly representative of the action and that nothing of any consequence could be accomplished

NOTES

without all those involved sitting around and chatting at length about their interminable family histories.

3   This particular device of dramatic resolution through divine intervention is most strongly associated with the work of Sophocles, but continues to find expression in modern literature, be it through the mysterious benefactor, or the hitherto unknown wealthy relative, or—my personal favorite—the sudden removal of a latex mask to reveal an entirely different set of characters and moral obligations altogether. For a fascinating study on the tensions within ancient Greek moral thinking, their expression in the medium of tragedy, and the incoherent inheritance of our contemporary situation, see Alasdair MacIntyre's *After Virtue* (Duckworth, 1981).

4   William Paley, *Natural Theology* (R. Fauler, 1802). The choice of a watch in Paley's analogy appears to be part of a long-running obsession with time-pieces in these sorts of arguments; well before the advent of the sort of wondrously complex mechanisms with which Paley was familiar, we find Cicero—the Roman statesman, orator, and philosopher—comparing the apparent design of the world with the intricate workings of a sundial or water clock. I suppose that it requires a particular type of personality, not to mention lifestyle, to find one's timepiece to be an inspiration for intellectual speculation, as opposed to merely an overbearing taskmaster. Sometimes I miss working at a university.

5   David Hume, *Dialogues Concerning Natural Religion* (London, 1779), Section VII. While Hume resisted the urge to populate his discussion with caricatures of his rivals and to generally humiliate his colleagues, the *Dialogues* were not published until after his death, on the advice of his friends and relatives who feared the consequences for his already scandalous reputation as an atheist and skeptic—although as usual, it was opposition from the academic community that eventually scuppered Hume's appointment to the University of Edinburgh rather than any direct interference from the Church. Nevertheless, Hume's *Dialogues Concerning Natural Religion* is without doubt a philosophical masterpiece, and over the course of little more than 100 highly readable pages manages to cover more arguments and with greater nuance than any of the interminable antireligious tracts that you will find wasting useful space in bookshops today. It is also out of copyright and free on your preferred e-reader, if that is your sort of thing.

6   Ibid., Section VIII.

7   Ibid., Section V.

8   This particular line of thought also finds its expression within traditional theology, and is generally referred to as *the problem of*

evil. The difficulty here is trying to reconcile the apparent existence of evil and suffering in the world with the existence of a deity who is both infinitely powerful (and thus capable of ending such suffering) and infinitely benevolent (and thus motivated to end such suffering). Clearly, the argument does not put any pressure on the idea of some all-powerful being in general—who may simply not care very much about our lives —but is seen as a challenge for those religions that believe in a more personally invested deity. The most popular response to the problem is to argue that some of the evil and suffering in the world is an inevitable consequence of some greater good, such as our capacity for freewill, and could not in that sense be prevented, no matter how powerful or caring the deity in question.

9 Jean-Paul Sartre, *La Nausée* (Librairie Gallimard, 1938), 184–5. This was the first work of philosophy that I ever read, and I vividly remember how powerfully it inspired me—even though I didn't have the slightest idea what it had inspired me *about*. After too many years of my copy collecting dust on the bookshelf, it gives me enormous pleasure to have found some way to shoehorn it into a serious argument.

Bibliography

Barnes, Jonathan (1984), *The Complete Works of Aristotle*. Princeton: Princeton University Press.

Chapman, M. (2008), *40 Days and 40 Nights: Darwin, Intelligent Design, God, Oxycontin, and Other Oddities on Trial in Pennsylvania*. New York: Harper Perennial.

Cooper, Lane (1935), *Aristotle, Galileo, and the Tower of Pisa*. Ithaca: Cornell University Press.

Copernicus, Nicolaus (1992 [1543]), *De Revolutionibus Orbium Caelestium*; translated by E. Rosen. Baltimore: The Johns Hopkins University Press.

Darwin, Charles (1859), *On the Origin of Species*. London: John Murray.

Davidson, Donald (1973), "On the Very Idea of a Conceptual Scheme," *Proceedings and Addresses of the American Philosophical Association* 47: 5–20.

Dicken, Paul (2016), *A Critical Introduction to Scientific Realism*. London: Bloomsbury.

Drake, Stillman (1980), *Galileo*. Oxford: Oxford University Press.

Frazer, James (1913), *The Golden Bough: A Study in Comparative Religion*. London: Macmillan.

Galilei, Galileo (2001 [1632]), *Dialogue Concerning the Two Chief World Systems*; translated by S. Drake. New York: Random House.

Ginsberg, Jeremy, Matthew H. Mohebbi, Rajan S. Patel, Lynette Brammer, Mark S. Smolinski, and Larry Brilliant (2009), "Detecting Influenza Epidemics Using Search Engine Query Data," *Nature* 457 (19 February):1012–14.

Hanson, N. R. (1958), *Patterns of Discovery*. Cambridge: Cambridge University Press.

Hayek, Friedrich (1944), *The Road to Serfdom*. Chicago: Chicago University Press.

Hempel, C. G. (1966), *Philosophy of Natural Science*. New York: Prentice-Hall.

Homer (1997 [c. 800 BC]), *The Iliad*; translated by R. Fagles. New York: Penguin Classics.

Hume, David (1975 [1748]), *Enquiries Concerning Human Understanding and Concerning the Principles of Morals*; edited by L. A. Selby-Bigge. Oxford: Oxford University Press.

Hume, David (1990 [1779]), *Dialogues Concerning Natural Religion*; edited by S. Tweyman. London: Penguin.
James, William (1890), *The Principles of Psychology*. Cambridge: Harvard University Press.
Jevons, William Stanley (1865), *The Coal Question: An Inquiry Concerning the Progress of the Nation, and the Probable Exhaustion of Our Coal-Mines*. London: Macmillan and Co.
Kahneman, Daniel (2011), *Thinking, Fast and Slow*. London: Penguin.
Kant, Immanuel (1998 [1781]), *Critique of Pure Reason*; edited by P. Guyer and A. Wood. Cambridge: Cambridge University Press.
Keynes, John Maynard (1951), *Essays in Biography*. New York: Horizon Press.
Kleinert, Andreas (1993), "Paul Weyland, der Berliner Einstein—Töter," in H. Albrecht (ed.) *Naturwissenschaft und Technik in der Geschichte, 25 Jahre Lehrstuhl für Geschichte der Naturwissenschaft und Technik am Historischen Institut der Universität Stuttgart*. Stuttgart: Verlag für Geschichte der Naturwissenschaft und Technik, 198–232.
Kuhn, Thomas (1957), *The Copernican Revolution*. Cambridge: Harvard University Press.
Kuhn, Thomas (1962), *The Structure of Scientific Revolutions*. Chicago: University of Chicago Press.
MacIntyre, Alisdair (1981), *After Virtue: A Study in Moral Theory*. London: Gerald Duckworth and Co.
Malthus, Thomas Robert (1798), *An Essay on the Principle of Population*. London: J. Johnson.
Newton, Isaac (2016 [1687]), *Philosophiae Naturalis Principia Mathematica*; translated by B. Cohen and A. Whitman. Oakland: University of California Press.
Paley, William (1802), *Natural Theology; or Evidence of the Existence and Attributes of the Deity*. London: R. Fauler.
Pennock, Robert and Ruse, Michael (eds) (2009), *But Is It Science? The Philosophical Question in the Creation/Evolution Controversy*. New York: Prometheus Books.
Plantinga, Alvin (2011), *Where the Conflict Really Lies: Science, Religion, and Naturalism*. New York: Oxford University Press.
Plato (1987 [c. 380 BC]), *Theaetetus*; translated by R. Waterfield. London: Penguin Classics.
Poincaré, Henri (1905), *Science and Hypothesis*. London: Walter Scott Publishing Company.
Popper, Karl (1945), *The Open Society and Its Enemies*. London: Routledge.
Popper, Karl (1957), *The Poverty of Historicism*. London: Routledge.
Popper, Karl (1959), *The Logic of Scientific Discovery*. London: Hutchinson.

Popper, Karl (1963), *Conjecture and Refutation*. London: Routledge.
Popper, Karl (1974), "Intellectual Autobiography," in P. A. Schlipp (ed.) *The Philosophy of Karl Popper*. London: Open Court Publishing.
Putnam, Hilary (1975), "What is Mathematical Truth?," *Philosophical Papers Vol. 1: Mathematics, Matter and Method*. Cambridge: Cambridge University Press.
Russell, Bertrand (1921), *The Analysis of Mind*. London: George Allen and Unwin.
Sartre, Jean-Paul (1938), *La Nausée*. Paris: Librairie Gallimard.
Smart, J. J. C. (1963), *Philosophy and Scientific Realism*. London: Routledge.
St. Augustine (1982 [c. 415 AD]), *The Literal Meaning of Genesis*; translated by John H. Taylor, *Ancient Christian Writers, Vol. 41–42*. New York: Newman Press.
van Fraassen, B. C. (1980), *The Scientific Image*. Oxford: Oxford University Press.
Wertham, Fredric (1954), *Seduction of the Innocent: The Influence of Comic Books on Today's Youth*. New York: Rinehard and Co.
Wittgenstein, Ludwig (1953), *Philosophical Investigations*. Oxford: Blackwell.

Index

Adler, Alfred 11–12 see also psychoanalysis
argument from design 152–63
Aristotle 26–30, 32–4, 37–43, 45, 64, 66–7, 72, 93, 108, 132, 144, 148–52, 155, 173
   and the Church 26, 45, 66–7
   ethics 148–9
   physics 26–30, 32–4, 37–43, 64, 66–7, 150–2
astrology xi, 173 see also pseudoscience

base-rate fallacy 86–9
Batman 86, 142
Big Data Revolution xii–xiii, 51–6, 61–3, 65, 67

causation 62–8
climate change xiii–xiv, 45–6, 123–7, 129, 131, 138–40
Copernicus, Nicolaus 33–7, 47, 67, 72, 108–9, 127, 141–4, 173–5 see also heliocentricism
creationism xi–xii, 5–9, 12, 17–18, 20–3, 78, 83, 91, 130–1, 159 see also argument from design; evolution

Darwin, Charles xiv, 83, 89, 155–9, 177 see also evolution
Davidson, Donald 114 see also relativism

Eddington, Arthur 10–11, 97 see also relativity, theory of
Einstein, Albert xiii, 10, 50, 84, 93–8, 100–2, 111–13, 126–7, 132, 136, 168–70, 175, 178–9 see also relativity, theory of
evolution xi–xii, xiv, 5–9, 19, 21–2, 77–8, 82–4, 89–91, 130, 154–9, 177 see also creationism; Kitzmiller v Dover; McLean v Arkansas

falsificationism 3, 9, 12–19, 59–60, 71–2, 103, 105, 109–10, 115, 155, 161, 179
First Amendment 6–8, 20
Freud, Sigmund 11–12 see also psychoanalysis

Galilei, Galileo xii, xv, 25–34, 38–47, 49, 64, 66–7, 72, 94, 109, 127, 129, 134, 143–4, 150, 156, 174–5
   Dialogue on the Two Chief World Systems 38–40, 45, 156
   Leaning Tower of Pisa 25–33, 38, 42
   trial xii, 40–6
   versus Aristotle 34, 47, 49, 67, 143–4, 150
Genesis, Book of 7, 42, 83 see also creationism
Google Flu Trends 51–2, 54–6,

INDEX

61, 63 *see also* Big Data Revolution

heliocentricism viii, 33–4, 37–42, 44–5, 47, 64, 77, 114–16, 142, 144, 173–6, 180
Hempel, Carl Gustav 179–80 *see also* induction; inference to the best explanation
Holmes, Sherlock 69–73, 77, 79, 85, 88–9, 91
homeopathy xi, 56, 66, 143 *see also* magic; pseudoscience
Homer 147, 149, 158
Hume, David xv, 58–9, 61, 63–4, 156–60, 175–6
    causation 63–4
    induction 58–9, 61
    natural religion 156–60

induction 57–62, 72–3 *see also* inference to the best explanation
inference to the best explanation 76–84, 87–8, 90–2 *see also* scientific success

James, William 32
Jevons, William Stanley 119–22, 125, 137–8

Kant, Immanuel 165–71, 176–7
Keynes, John Maynard 121–2, 125, 137
Kitzmiller v Dover 20–1, 23
Kubrick, Stanley 1, 103
Kuhn, Thomas 107–12, 114–15, 180

Leibniz, Gottlieb 48, 67
logical positivism 169–71, 180
luminiferous aether 98–102, 111, 126, 178 *see also* relativity, theory of

magic 65–7
marxism 10–11 *see also* pseudoscience
Maxwell, James Clerk 94, 98–9, 126, 177–8
McLean v Arkansas 5–9, 17–18, 20–1, 23
moon landings 1–3, 15–17, 103

Neptune, discovery of 15–16
Newton, Isaac 3, 15–17, 47–52, 56–7, 84, 93–4, 108, 115, 126–7, 129, 132, 143, 165–8, 170, 175
    lack of data 49–52, 56–7, 61–2
    physics 3, 15–17, 50–1, 56–7, 84, 94, 115, 126–7, 132, 143, 166, 168, 170
    self-experimentation 49–50, 64
Nixon, Richard xvi, 1–2, 15, 17, 131, 133

Paley, William 152–3, 155, 159
Plantinga, Alvin 90–1
Plato 148–9, 152, 155, 159, 173
Poincaré, Henri 127, 178
Popper, Karl 3–5, 9, 10–12, 15, 18–19, 24, 59–60, 71, 72, 97, 107, 109–10, 134–40, 179
    on evolution 18–19
    on induction 59–60
    intellectual background 10–12
    science and democracy 4–5, 24, 134–40
    see also falsificationism
postmodernism 98, 162, 181 *see also* relativism
pseudoscience ix, 9–12, 18, 21–2, 59, 75, 103–4, 143, 165
psychoanalysis xi, 11–12, 143 *see also* pseudoscience
Ptolemy, Claudius 34–8, 72, 77, 108, 116, 142–4, 173–4

ptolemaic system 34–7, 77, 108, 116, 142–4
see also Copernicus; heliocentricism
Putnam, Hilary 80–1, 90, 189 n.4

quantum mechanics viii, xii, 18, 64, 126, 178–80

reasoning, heuristics 84–91 see also base-rate fallacy
relativism 112, 114–17
relativity, theory of xiii, 10, 11, 50, 84, 93–8, 100–2, 111–13, 126–7, 132, 136, 168–70, 175, 178–9
 Michelson-Morley experiment 99–100
 public reception xiii, 93–8, 111, 113

Sartre, Jean-Paul 162
Scheiner, Christoph 43, 45
scientific failure vii–ix, 82–4, 123–7, 129–32
scientific revolution 20–1, 38, 47–9, 66–7, 141–6
scientific success vii–ix, 79–84, 87–8, 90–2, 171–2
Scopes Monkey Trial 6 see also creationism; evolution
Seagal, Steven 131
Semmelweiss, Ignaz 73, 74, 75, 76, 77, 79
Smart, J. J. C. 79–80, 84, 90
Socrates 70, 112, 114
St Augustine 42–3
Superman 86, 142

Thales of Miletus 144–9, 152, 153–5, 158

Van Damme, Jean-Claude vii, 14, 40, 55, 86, 101, 131, 142, 180–1
Van Fraassen, Bas C. 83, 87

Weyland, Paul 95–8, 102–3, 111–13
Wittgenstein, Ludwig 106